AF557838

PEOPLE FIRST

'As Secretary, Ministry of Housing & Urban Affairs, Government of India, I keenly noticed the enthusiasm, commitment and passion with which Mathi, in his capacity as Principal Secretary, Housing & Urban Development Department, Government of Odisha, pushed the national agenda for providing clean drinking water in all urban local bodies in his water-scarce state. Executing 24×7 supply with the quality of *Drink from Tap* in the city of Puri was a culmination of his efforts. He made this impossible-looking target possible in the holy city and many other towns.

'His narrative, experiences and learnings in this voyage will inspire many civil servants and experts in the field. *People First* shows how this journey was not only about a technological solution but was also about engaging and empowering people, especially women, in socio-engineering transformation. It is a must-read for practitioners in the water sector.'

—Durga Shanker Mishra
Chief Secretary, Uttar Pradesh

'This book by G. Mathi Vathanan is a ray of hope. Mathi describes in great detail his journey over a seven-year period with all the accompanying trials, tribulations as well as successes. This book is a valuable addition to the knowledge about urban water provision not just in India but globally. It provides insights that any policymaker, public servant or development practitioner would find useful. His empathy for urban low-income families and, in particular, recognition of the invaluable contribution of migrants to our cities is best illustrated in his words: "Not only do they create the city, but they also run the city". This emphasis on addressing inequities with respect to access to water supply enhances the contribution of the

Drink from Tap effort in Odisha. The book dispels several myths about the feasibility of imagining 24×7 water supply provision in urban India and is a reminder that "investment in water is nothing but an investment in health".'

—Sumit Bose
Chairperson, WaterAid India

'In his book on Odisha's transformative journey in urban water supply management, Mathi delves into the state's groundbreaking strategies that not only revolutionized its water infrastructure but also positively impacted the lives of its residents, especially women. The narrative showcases Odisha as a model of community engagement, sustainability and inclusivity.

'Central to the book's theme is the fundamental aspect of gender parity, public health and climate resilience within water supply management. Odisha's story goes beyond achieving universal drinking water access, shedding light on its empowering effects on women, liberating them from poverty and enabling pivotal roles within their communities.

'The book emphasizes the synergy of community partnerships in ensuring water security while simultaneously creating economic opportunities. The innovative approach by Jal Sathis in addressing complex challenges related to water supply distribution and consumer management, with a focus on sustainability and consumer-centricity, is truly inspiring. There is a notable shift towards viewing public water supply beyond mere infrastructure—as a vehicle for trust-building and long-term sustainability.

'In essence, this book serves as evidence of the transformative impact of strong leadership, innovative policy reforms and community partnerships. It provides valuable insights and lessons for policymakers and practitioners working towards sustainable development through urban and rural water supply management.

'By shining a spotlight on Odisha's success, *People First* serves as a blueprint for other regions striving to ensure universal access to clean, safe drinking water.'

—Rodger Voorhies,
President, Global Growth & Opportunity Division,
Bill & Melinda Gates Foundation

'This book captures a remarkable vision—one that is of huge relevance to water and sanitation needs around the world. Mathi's vision is inclusive, seeking to transform the lives of those lacking the most basic of services. It also challenges convention on which solutions to pursue. Moreover, Mathi shows through his experiences that the change needed to implement this vision is possible. What is truly outstanding is the extent to which he has been a driving force in the different dimensions needed to bring about this change from policy and engaging political leaders to community involvement and addressing real-world issues around project contracts and scaling-up solutions.'

—Kalanithy Vairavamoorthy
Executive Director, International Water Association

'The moral of this great story is that it takes one strong person to declare a situation unacceptable and hence set oneself up to transform it. Upending the status quo, particularly in cases involving severe water shortages, high water losses, etc., is not simple, nor is it just an engineering job; it requires mass mobilization, community-wide behavioural shifts and, above all, resolute autonomy coupled with dedicated support from the Government. Mathi not only achieved this but secured unwavering backing from his Chief Minister and all key stakeholders of the programme.

'*People First* is a source of inspiration and significantly contributes to the growing body of work on revitalizing service delivery, not only in economically challenged regions but also in the burgeoning middle-income economies worldwide.'

—Dr William Muhairwe
Executive Director, Global Water Leaders Group (UK) and Former CEO of National Water and Sewerage Corporation (Uganda)

'Mathi Vathanan's book is a remarkable story of unparalleled urban transformation in the history of developing India. He tells a good story of how sensible ideas, pilot-tested and scaled up, got the support of the political establishment, allowing him to drive an indigenously developed program that enables urban Odias to quench their thirst safely. Mathi is a different kind of civil servant. His book tells us not just about his vision, commitment and results but also about the ways in which he translated his passion into building teams and inspiring them. While his humility will not let him say this, it will be his substantive, permanent legacy. This book is the story of that passion and the legacy he so zealously created.'

—Arjun Thapan

Chair, Board of Trustees, WaterLinks, Philippines

'If you care about providing clean drinking water to all urban households, "24×7" with no slum-dweller left behind; if you wonder about the viability of ensuring continuous water supply in a middle-income climate-challenged country, especially when so many of its own professionals think it impossible—then read this book! If you want practical solutions, from ferrules to Jal Sathis, as well as the inside story from a leader who is in a direct line of the greatest public health enablers, Mathi Vathanan's explanation of how he and his colleagues have been delivering 24×7 water to Odisha in less than a decade is inspiring.'

—Dr Richard Franceys

WEDC-Loughborough, IHE Delft & Cranfield University, UK

'There is something for everyone in this book, including technocrats, administrators, social scientists and even politicians. It is not an exaggeration to say that this book is a Holy Grail on drinking water services—an inspiring read for all those associated directly or indirectly with such a service. This book has an answer to all the problems associated with such a humongous task. I congratulate Mathi for his highly commendable contribution and wish him the very best in his future endeavours.'

—M.S. Mohan Kumar

Former Professor, Water Resources & Environmental Engineering, Indian Institute of Science, Bangalore

PEOPLE FIRST

How Odisha's *Drink from Tap* Mission Quenched Every Thirst

G. MATHI VATHANAN

RUPA

First published by
Rupa Publications India Pvt. Ltd 2024
7/16, Ansari Road, Daryaganj
New Delhi 110002

Sales centres:
Bengaluru Chennai
Hyderabad Jaipur Kathmandu
Kolkata Mumbai Prayagraj

P-ISBN: 978-93-90260-93-5
E-ISBN: 978-93-90260-97-3

Second impression 2024

10 9 8 7 6 5 4 3 2

Printed in India

Dedicated to
the cause of providing safe, clean
and affordable drinking water to all,
including the underprivileged in urban areas

CONTENTS

FOREWORD

Ensuring access to safe drinking water is vital for human health and dignity. G. Mathi Vathanan's book provides a compelling account of his journey in the water sector and the transformative change he helped bring in Odisha's water system. This is a captivating insider's account of a number of developments, including Puri City's pioneering effort to provide its residents with a round-the-clock water supply.

This book is not only a testimony to Mathi's passion and commitment to water-related issues but also a great lesson in leadership skills and how to drive ambitious initiatives. It describes how Mathi and his team conceptualized and undertook some of the most ambitious water initiatives in the state, dealing with hurdles along the way. This includes challenges with the bidding process, setbacks due to COVID-19, other challenges that made the process complex and frustrating at times as well as how Mathi and his team worked tirelessly to overcome these issues.

Mathi's book is a timely reminder of the critical importance of ensuring access to safe drinking water, particularly for children. My co-authors—Steve Luby, Brandon Tan, Ricardo Martens and Witold Wiecek—and I recently released a meta-analysis by combining data from 15 studies, which found that water treatment reduces all-cause child mortality by a quarter. This suggests that water treatment is one of the most cost-effective child health interventions available.

Having worked on water treatment for two decades now, I've found Odisha's achievements in expanding access to piped water under the leadership of Mr Mathi to be hugely exciting and inspiring. In early 2023, I had the opportunity to visit Odisha and see their *Drink from Tap* programme.

During my visit, I also witnessed the community partnership model in urban water supply management and heard how this model helps bridge the gap between the water supply authorities and consumers. Mathi explains in the book that the community partners, or Jal Sathis, are pillars of the urban drinking water management system in Odisha.

One lesson from this programme is that the public sector has an important role in the provision of water. In some quarters, there has been pessimism about the ability of the public sector in middle-income countries to provide quality services. The successes of the *Drink from Tap* programme so far have generated optimism that the public sector can deliver water access effectively, thereby making it a model not only for India but also for the world.

—Michael Kremer
Professor of Economics and Public Policy
at the University of Chicago
Recipient of the Nobel Prize in Economics (2019)

FOREWORD

Water makes or breaks any society; it is about health, well-being and livelihoods. In an age of climate change with increasing rainfall variability, supplying water to all will become a greater challenge.

Today, cities are still finding it difficult to supply water to everyone, which leads to the over-extraction of groundwater as people have no option but to secure their own supply. Furthermore, water supply is erratic, which leads to pipeline contamination, so polluted and unclean water flows from the taps.

This is why this ground-level and transformational work by G. Mathi Vathanan needs to be understood, celebrated and practised. His water journey shows that it is possible to fulfil citizens' right to drink clean and safe water straight from the tap.

Ultimately, water wisdom is about building an inclusive paradigm that is affordable and sustainable. This book of experience and practice shows what can be done and how.

—Sunita Narain
Director General, Centre for Science and Environment
Padma Shri Awardee

PREFACE

Not everybody is privileged enough to gain entry into the Indian Administrative Service (IAS) and serve as an IAS officer. In my opinion, the IAS offers the most opportunities to make a difference in people's lives. Working in the Odisha cadre, under the leadership of Honourable Chief Minister Naveen Patnaik has been a blessing for me. The freedom with which an IAS officer gets to work in Odisha is unparalleled, and the kind of platform that Odisha provides to think, conceptualize and implement people-centric programmes without any undesirable political interference is unmatched by any other state or union territory in the country.

Therefore, it has become imperative for me to make the best use of this once-in-a-lifetime opportunity to address one of the biggest challenges faced by the state: to make clean and safe piped water accessible to every home in urban Odisha.

Odisha's capital city of Bhubaneswar became the first million-plus city in the country to achieve 100 per cent pipe water house connections, followed by 105 out of the state's 115 cities. Thereafter, Puri became the first city in the country to deliver 24×7 tap water of IS 10500 quality standards on 26 July 2021. This achievement was soon replicated by 23 other cities. It was then that I felt that the most defining moment of my professional life had arrived.

The recipe for the water supply transformation achieved in urban Odisha has several ingredients, starting from the political will of the top leadership to community participation at the grassroots level in addition to human and financial resources, policies, programmes, schemes, etc. The objective of this book is to mix all these ingredients along with the re-engineered processes and learnings and make them available for those who wish to emulate or better the Odisha model of 24×7 urban water supply systems.

Left with about two years to retire and a professional team so passionately mentored by me through the years, which was motivated to achieve seemingly unachievable goals in public service with zeal and ease, I now stand at the threshold of the large-scale implementation of the Drink from Tap (DFT) mission in urban Odisha along with Chennai, where we have been engaged as project management consultants. I have to forge ahead with speed and scale to accomplish the formidable task before I sign off as the Additional Chief Secretary of the Housing and Urban Development Department (H&UDD) of Odisha.

Now, as I reminisce about the years spent on this journey, I strongly believe that there existed a driving force within me, other than my official responsibility, guiding me along the way. While writing this book, I revisited my past and present works in Odisha, thereby unravelling the magical transformations that my driving force made me envision, conceptualize and bring into existence.

This book is an attempt to narrate my memories of those significant moments. I would be glad if this book succeeds in inspiring a larger audience and our experiences are used by water leaders at the city, state and national levels.

INTRODUCTION

Urban dwellers have always faced the brunt of water challenges. Their anguish has remained enduring, persistent and often overwhelming. The recent decade has seen extraordinary apprehension among common people regarding water. Indisputably, water-related issues and crises are mostly due to the unsustainable management of water resources. According to a recent estimate, half of India's population will be living in urban areas and will face acute water problems by 2050. For urban Odisha too, water crises may become irredeemable. A comprehensive understanding of human-induced water scarcity and potential solutions for towns and cities is urgently required to promote a more sustainable and liveable urban future.

Predominantly, public water supply schemes are globally considered an engineering endeavour. However, during my journey and due to my deep association with the water sector, I have realized that this is more of a socio-engineering venture.

For the sustainable and successful implementation of such public water supply schemes, other than the key engineering components, we certainly require strong political will, proactive administrative actions, a propitious work environment, and community support and partnership. All of these ingredients contribute equally.

When we embarked on this journey, the water supply system in Odisha:

- Was intermittent in nature and prone to contamination, which often led to public health hazards.

- Had large demand and supply gaps and was characterized by sporadic water supply, inequities in supply, inconsistent pressure and high losses due to leakage.
- Covered only 30 per cent of the households with piped water connections, hence leaving the rest to depend on public stand posts and other sources or spend long hours in serpentine queues to obtain drinking water.
- Was plagued by consumer distrust and dissatisfaction due to poor service delivery and a lack of community connect in water supply management.

The system underwent transformation through several water supply missions and projects steered by the Government of Odisha during my tenure as the secretary of the H&UDD since 2014.

Our team's learnings are highlighted in this book, which outlines the maze of perspectives, processes, propositions and practices that we had to innovate, improve or redefine. The narratives also underline the importance of belief systems, moral values, the team's commitment and, above all, the political will of the leader at the helm of state affairs that made the transformation possible.

We conceived the journey to pursue the shared dream of the state's leadership by defining the destination, which is dynamic according to changing aspirations. Because the entire journey was scripted by us, our team worked with passionate commitment to ensure success with the firm belief that failure was not an option.

This book recounts my diverse experiences during the adventurous and tumultuous transformative journey of Odisha's cities from being water-scarce to water-secure. I have focused on significant engineering aspects and other social aspects, which made this unique model unparalleled but replicable across the country.

1

THE DAWN OF MY RESOLVE

The seeds of a strong and deep commitment to ensuring a water-secure urban Odisha were sown in me early in my career during my second posting as a young IAS officer in 1997 as a sub-collector. That was when Berhampur, Odisha's third-largest city, was facing a grave water crisis.

Back then, I was a helpless witness to the suffering that the crisis had caused to the people, thus leading to public unrest, law and order situations and consequent assaults on the officials of the Public Health Engineering Organization (PHEO).[1] The state had no choice but to bring water by train wagons from Visakhapatnam in the neighbouring state of Andhra Pradesh, 250 km away, to tide over the city's water crisis during that summer.

Challenged by the unrelenting crisis and unable to cope with the public pressure, the then executive officer (EO) of Berhampur Municipality fled the city overnight. As an immediate response to overcome the situation, I was given additional charge as the EO of the municipality, and my first job as the EO was to ensure the availability of drinking water for people in all areas of the city. Soon, I realized that besides supplying water to homes, I was responsible for comforting

[1]PHEO was formed in 1956 by the Government of Odisha under its Housing & Urban Development Department (H&UDD) to provide water supply and sewerage facilities to the urban local bodies (ULBs) of Odisha.

the people suffering from the water crisis for a long time.

I started by regularly visiting the affected areas, talking to people and reassuring them. The experience gained during that period by dealing with an acute water crisis greatly influenced my thought process in dealing with people-centric problems and instilled a strong resolve in me to ensure piped water supply to every home in urban Odisha.

When I went to call on the revenue divisional commissioner (RDC) within a few days of joining in Berhampur, I found him surrounded and being heckled by local advocates for not solving the week-long water crisis that the city had been experiencing. The RDC was trying hard to calm the hecklers down. By explaining that his own residence and family had also gone without water for a few days, he hoped to neutralize their rage and gain sympathy.

Contrary to his hopes, his statements only worsened the situation, as the advocates became more enraged and agitated over what they perceived to be the helplessness of the RDC, who was heading the division with 10 districts under his command, in addressing the water crisis. One advocate could be heard shouting as to how the RDC, who was incapable of getting water for his own residence, could ensure water for others. This incident was another lesson for me regarding what not to say while dealing with such crises.

The extreme water crisis and resulting hardships suffered by the people of Berhampur triggered my deep resolve to give top priority to water supply later in my career. This incident resurfaced in my mind when I assumed charge as the secretary of the Housing and Urban Development Department (H&UDD) with drinking water management as one of my core responsibilities.

Twenty-two years later, in March 2019, I was felt elated when Odisha's first mega urban water supply scheme for Berhampur was inaugurated by the chief minister (CM) to permanently solve the city's perpetual water crisis.

In my address during the inaugural function, I recalled my personal connect with the water supply crisis faced by the city 22 years ago. I can say that my role, right from the conceptualization stage to the completion and inauguration of the project, was one of the most satisfying aspects of my career.

After taking charge of the department, I strongly felt that drinking water issues in urban areas should become my top priority. With only 40 per cent of urban areas being covered by piped water supply networks and less than a third of the population in urban areas having house connections (HCs), the water supply situation was grim. The total water supply shortfall was 150 million litres per day (mld) at the state level.

The average water supply at that time was 60 litres per capita per day (lpcd) against the national benchmark of 135 lpcd[2] (about 13 buckets with a bucket capacity of 10 litres, which are normally used in our bathrooms). To me, this appeared to be an incorrect depiction as the estimated supply of 60 lpcd was calculated based on the entire city's population instead of the population in the piped network area.

This estimate meant that more water was being provided to the 40 per cent of areas covered by the pipe network, whereas the remaining 60 per cent of areas were fed only by hand pump tube wells or polyvinyl chloride (PVC) tanks filled by water tankers. For example, in Bhubaneswar, the quantity of water supplied in 2015 was 350 lpcd against the national benchmark of 135 lpcd (about 35 buckets compared with the national benchmark of 13 buckets of 10 litre bucket capacity).

The 350 lpcd of water supply was estimated considering

---

[2]As per the Ministry of Housing and Urban Affairs (MoHUA), Government of India, 135 litres per capita per day has been suggested as the benchmark for urban water supply.

the entire population of Bhubaneswar, as the population covered by the pipe network (in 40 per cent areas) received about 600–700 lpcd (about 60–70 buckets per person or 300–350 buckets per family); the remaining 60 per cent of households were left to get by with hand pump tube wells or PVC tanks.

In such scenarios, people have no option other than to fend for themselves. The persistent scarcity and irregular availability of water in non-networked areas help the private water businesses to flourish.

Thus, while some areas in Odisha's 105 cities were being provided with a very high volume of water, the remaining non-networked areas received hardly any water supply. For instance, Bhubaneswar was being supplied with a high volume of 350 lpcd, whereas Belpahar, a faraway small town, was getting only 11 lpcd (one bucket of water per person per day).

Out of the 105 cities, the ones that were bigger and closer to the state capital were getting higher per capita water supply, whereas water supply in cities that were farthest from the state capital, mostly in western Odisha, was paltry. The inequities that existed within and among the cities were extremely high.

The urban water supply situation in Odisha in 2015 brought back my memories of Berhampur. When these water supply precarities and system disparities were presented to CM Naveen Patnaik, his face became very grim. He turned towards me and, in a stoic manner, said, 'We must do something,' and left the meeting. This moment lingered in my thoughts from then onwards and propelled me to prioritize and resolve these issues throughout the state as early as possible.

THE ESSENCE

The journey of a thousand miles begins with one step.

—LAO TZU

Throughout history, many of humanity's greatest accomplishments have emerged from times of crisis. A crisis can be seen as an opportunity for growth and change, as it often brings to light the deep-seated issues within our society and inspires us to take action. The water crisis in Berhampur was not just about scarcity; it was a reflection of the deeper malaise in our society.

Dealing with the people's wrath as the sub-collector of Berhampur and as the acting EO of the municipality gave me first-hand experience with the severity of the crisis and the problems it caused.

I have learned that one truly understands the magnitude of the issue only when one steps into the sufferer's shoes. When we prioritize the needs of the community, the solutions we derive are effective, equitable and sustainable in the long run. Along with providing safe drinking water, the state needed a people-centric approach and a comprehensive strategy for sustainable water management.

The water scarcity crisis that I experienced in Berhampur sparked a sublime resolve within me.

In the face of adversity, we are capable of great strength and endurance.

2

THE RISE OF THE BEGINNING

On 25 June 2015, the Government of India launched four national flagship missions for the urban sector: Atal Mission for Rejuvenation and Urban Transformation (AMRUT), Smart Cities Mission (SCM),[3] Pradhan Mantri Awas Yojana-Urban (PMAY-U) and National Urban Livelihood Mission (NULM). Of these, AMRUT focuses on improving core urban infrastructure, including piped water supply, storm water drainage, sewer/septage management and urban transport. It introduced a new approach by stipulating the national hierarchy of priorities to be followed in selecting projects; drinking water was the topmost national priority, followed by sewerage/septage system, storm water drainage system and public transport.

This was perhaps the first time that the principle of a national hierarchy for sectoral priorities was introduced at the national level for a scheme/mission. Generally, cities and states are given the flexibility to select projects from a list of sectors, such as water, road and drainage, for implementation according to perceived local needs and priorities.

The national hierarchy of priorities stipulated by AMRUT is non-negotiable, which means that the funds provided under AMRUT must necessarily be spent in the selected cities for

---

[3]Smart Cities Mission, Ministry of Urban Development, https://smartcities.gov.in/about-the-mission. Accessed on 20 December 2023.

the selected schemes in the stipulated order of priority. Until all homes in all AMRUT cities are provided with piped water supply connections, the scheme funds cannot be utilized for any other purpose in any AMRUT city.

This novel and unique aspect of the scheme fascinated me the most because it has given the absolute mandate and authority to implement the piped water supply scheme in all the nine[4] AMRUT cities to achieve 100 per cent HCs without allowing the diversion of funds for any other purpose in any other city.

However, it also created challenges in handling the priorities of the elected municipal councils when they were not in sync with national priorities. For instance, the city council of Bhadrak, one of the nine AMRUT cities, resolved to prioritize drainage projects instead of water supply despite having only 50 per cent of the city covered with pipe networks.

Influential and affluent locals who already had piped water supply at their homes were not interested in utilizing AMRUT funds to provide piped water supply to the non-networked areas and the underprivileged class (mostly the urban poor and slum dwellers)

Without adhering to AMRUT guidelines and the state's directive, a resolution was passed in the municipal council meeting for utilizing the scheme funds for works other than water supply, claiming it to be the collective resolve of the city as drinking water was not a priority.

The national hierarchy of priorities mandated by the scheme came in handy for us to reject the municipal council resolution.

We issued an advisory to the municipal council to reconsider and utilize the funds to provide water supply to

---

[4]Balasore, Baripada, Berhampur, Bhadrak, Bhubaneswar, Cuttack, Puri, Rourkela and Sambalpur.

all the non-networked houses. It was conveyed that funds earmarked for the city would be diverted to other cities if not utilized as per the AMRUT guidelines.

Our firm stance and threat of diversion of funds to other cities compelled the Bhadrak Municipal Council to reverse their decision and take up water supply projects.

I saw the top priority given to piped water supply by AMRUT as a golden opportunity to convert and realize the long-cherished dream of permanently resolving Berhampur's age-old drinking water struggle, which had been bothering the CM since his early days in politics, which I had also witnessed since my sub-collector days in 1997.

Berhampur, an important commercial, educational and cultural hub in southern Odisha, was marred by the scarcity of water and a deficient distribution system. The city's notoriety for its water woes consistently gained public attention. The city's population is approximately 4 lakh and its domestic water demand is 60 mld.

During summers, the demand rises to 70 mld and supply falls to 35 mld. Factors such as drying of riverbeds, dwindling water volume in the Rushikulya river during summers, illegal water tapping by households via pumps, and pipeline leakages due to the poor maintenance of the pipe network lead to the reduction in water supply.

These conditions further aggravated the already stressed supply system and made the situation precarious. The PHEO had no option but to supply water through tankers and the situation worsened during the peak of summer.

As a comprehensive response and permanent solution to the city's water crisis, the Greater Berhampur Water Supply Project (GBWSP) was launched in 2016. With an outlay of ₹4,900 million to bring water from the Janibili anicut in the Rushikulya river, 55 km away from the city, the project was to be completed within an ambitious period of two years.

Right from tendering, the project faced several obstacles and challenges. The lowest bid received was 17 per cent less than the indicative cost, and as per the guidelines issued by the finance department (FD), any bid with a –14.99 per cent tender cost was to be considered rogue and treated as invalid.

Though the standard bidding document (SBD)—which had the approval of the FD—stipulated that the lowest bid would be approved, the Berhampur bid was not approved by the FD because the quoted price was less by more than 14.99 per cent.

After a detailed analysis of the bid document, we discovered that the tender was according to the engineering, procurement, construction (EPC) mode based on an indicative cost and not the estimated cost. To our dismay, we noticed that the PHEO had erroneously recorded the indicative cost as the estimated cost in the file, which led to the rejection of the lowest bid by the FD.

Though the stipulation of treating –14.99 per cent bids as invalid bids applied to turnkey projects or item rate contracts and not to EPC contracts, the FD was still reluctant to understand the logic and referred the issue to the works department for its views, which in turn responded with ambiguous remarks.

I was very clear that our interpretation was right, so I didn't let go of the tender and insisted that our case be taken to the Cabinet for the approval of our stance and the tender.

The Cabinet not only appreciated our viewpoint and approved the tender but also instructed the works department to make appropriate amendments to the Works Code by incorporating modalities for processing EPC tenders. This Cabinet approval helped us avoid a possible delay of about one year in the retendering process and was the first successful step taken in the implementation of the project.

Berhampur's water demand was earlier met with water drawn from the Rushikulya riverbed in Badamadhapur, treated at the Kaliabali water treatment plant (WTP) and transported through a 40-km-long pipeline. The supply was facing frequent interruptions as the villagers along the supply line resorted to illegal tapping and breaking of pipes at several places to meet their water requirements, including irrigation for their crops. Guarding the 20-km-long vulnerable stretch with the help of local police and PHEO officials had been impractical and unsuccessful since 1997.

I realized that it was unreasonable to expect the villagers to provide peaceful and undisturbed passage for water transported for Berhampur through pipelines crossing their villages when they themselves were suffering from highly inadequate supply with limited access to water through hand pump tube wells. My water crisis experience from 1997 gave me the insight that a more inclusive approach was required to meet the villagers' genuine needs.

Thus, to ensure a win–win arrangement, we designed the Janibili project as the state's first composite project for urban and rural areas to ensure adequate water supply to 53 villages between Janibili and Berhampur.

The success of the idea saw the project mature into an everlasting solution to the water crisis faced by Berhampur and all the nearby villages through which the pipelines passed.

It has also become a trendsetter in urban–rural convergence for inclusive service delivery. Recently, the state's Panchayati Raj and Drinking Water (PR&DW) Department has also been floating several mega piped water supply projects catering to rural and urban areas, thus achieving convergence.

Due to innumerable bottlenecks and hurdles, project execution was running behind schedule. A year and a half after the commencement of the project, the situation was alarming; a third of the two-year project period had already elapsed and a substantial portion of the work remained to be completed.

The CM, also a five-time legislator from the district, had promised the people of Berhampur that the project would be completed before March 2019. We had to make it happen in due time to fulfil the promise made by the CM.

With elections to the state assembly fast approaching, it was an extremely horrifying situation for the team as we firmly believed that failure was not an option for us; we could not afford to let the CM down in the district of his own constituency.

We did not see this as a simple political agenda of the CM but as his genuine, long-nourished dream for the place he was emotionally attached to. Mr V. K. Pandian, IAS, his then private secretary (PS) and now Chairman of 5T initiatives, was monitoring the progress very closely. On many occasions, I had to seek his intervention to sort out issues on administrative and political fronts.

This project was special because it was the CM's dream, which was implemented by the team under my leadership and steered by Pandian; both of us had first-hand experience in dealing with water crises.

Realizing the urgency and sensitivity of the initiative, we had to quickly come up with innovative strategies to compress the execution period for the remaining work, turn things around and complete the project within the next six months.

Having seen the repeated inability of the executing agency to catch up with the shortfalls in targets and rising to the occasion to complete the project within the stipulated time while seeing our hopes slowly fade away, I consulted Mr Pandian.

We decided to bring in Mr Pradipta Kumar Swain, a superintending engineer (SE) who had earlier successfully delivered the Bhubaneswar Bulk Water Supply Project.[5] He was made the officer on special duty (OSD) for the project.

More than 10 local contractors from various districts within the state were mobilized, and they came with their human resources and machinery. The balance work was redistributed amongst these contractors along with the main executing agency by invoking the relevant contractual clause at the employer's discretion for completing the delayed work at the risk and cost of the main executing agency. These teams simultaneously engaged on all work fronts that were behind schedule. It was a warlike situation with the entire functional set-up engaged in round-the-clock execution.

The project was completed on time, and it was a happy moment when the CM dedicated the project to the people of Berhampur on 3 March 2019, along with 200 other projects across the state. The promise was fulfilled, thanks to the strong political leadership and the committed and collective effort of my team.

Implementing the state's first mega water supply project for Berhampur after braving innumerable challenges led to great learnings besides the formulation of reformative policies and implementation measures. It strengthened our commitment to implement similar projects in other cities in Odisha. I shall always reminisce with pride that the Berhampur project was the launch pad and will remain a milestone in my water journey.

Timely land acquisition is a key factor for the completion of projects within the scheduled time. We have made it a practice to ensure that land acquisition for WTPs, elevated storage reservoirs (ESRs) etc., which is required for the project, is a prerequisite for the finalization of tenders.

---

[5]The Bhubaneswar Bulk Water Supply Project was the PHEO's first public–private partnership project to supply water to Khordha and Jatni and nearby areas.

However, for pipe-laying work, the local public often stops the work by claiming ownership of the portion of land where the pipe is being laid. This leads to unforeseen delays and uncertainty in project completion. In the linear component of infrastructure projects, such as pipelines, land acquisition of narrow strips of land from land parcels is not easy.

The primary reasons are that, one, owners hesitate to spare a long strip or portion of land from their land parcel and, two, complete land acquisition is not a viable proposition because after laying the pipelines, the land need not be used further and it can be returned to the landowner for cultivation or other use.

It also causes delays in land acquisition, which ultimately delays project implementation. To mitigate the issue, we came up with a policy on the Right of Use of Private Lands for Laying of Water Supply Pipelines[6] along the lines of the Right of Use of Private Lands for electrical infrastructure.

Under the policy, the PHEO enters into an agreement with the owner of the private land to obtain the Right of Use of Land and registers it with the concerned registering authority. The PHEO would compensate the owner for the acquisition of the Right of Use of Land by paying up to 25 per cent of the prevalent market value of the land, an additional payment of an equivalent amount as solatium and the full value of the damage caused to standing crops and structures on the land.

The policy was rolled out in November 2017, and private land acquisition issues related to laying pipelines were resolved, which would otherwise have taken a lot of time in the regular acquisition process and would have resulted in a substantial increase in project cost.

---

[6]H&UDD, Government of Odisha, Notification No. 27498 dated 25 November 2017.

This is the first-of-its-kind step in the entire state for similar infrastructure projects. Along with the GBWSP, we implemented the BBWSP using the public–private partnership (PPP) model as a first in the country.

When the Indian Institute of Technology (IIT) Bhubaneswar and National Institute of Science Education and Research (NISER) were proposing to move to their permanent campus in Jatani on the outskirts of Bhubaneswar, it became imperative for the state to implement a bulk water supply project to cater to the needs of these two premier institutions.

This project was conceptualized as a PPP project as it involved a substantial investment. For the project to be financially viable and attractive to private investors, the project size had to be reasonably large.

Therefore, we included the Odisha Industrial Infrastructure Development Corporation's (IDCO) industrial estates and nearby Khordha and Jatani municipalities within the scope of the project. This project has several unique distinctions:

- The country's first PPP project in the water supply sector covering all components of the water supply system and catering to domestic, institutional, industrial and commercial usage needs.
- The only project in Odisha for which environmental, forest, State Wildlife Board and Central Wildlife Board clearances were obtained.
- The first PPP project in Odisha for which Central Viability and State Viability Gap Funding (VGF) were availed.

It would not be an overstatement to say that the first-hand experiences gained during the implementation of the AMRUT projects, especially the Berhampur water supply project and BBWSP, gave our team the confidence and impetus to dream big and march ahead.

THE ESSENCE

Opportunities are like sunrises. If you wait too long, you miss them.

—WILLIAM ARTHUR WARD

In the world of development, opportunities are rare and one must be ready to seize them with unbridled passion. For us, that opportunity arrived when we started implementing the AMRUT scheme. This was a watershed moment as drinking water projects were at the top of the national hierarchy of priorities, which was non-negotiable and prohibited the diversion of funds for any other purpose before achieving universal piped water supply in the cities covered under AMRUT.

We harnessed the priority accorded to drinking water supply by AMRUT to bring piped water supply to all the nine AMRUT cities in the state. Seizing the opportunity, we developed a complementary state scheme, Buxi Jagabandhu Assured Water Supply to Habitations (BASUDHA), thereby prioritizing drinking water supply for the remaining 105 cities of the state. Every noble beginning has the potential to lead to something extraordinary, and it is up to us to seize the opportunity and make a difference.

3

SLUMS: THE UNSEEN THIRST

As per the United Nations (UN) definition, a slum is a place which lacks access to improved water, sanitation, sufficient living areas and durable housing. With rapid urbanization, it is estimated that globally, around 3 billion people will be living in slums by 2030. Currently, more than 60 per cent of the world's slum population is living in Asia. About 20 per cent of India's urban population lives in slums.

An estimated 25 per cent of Odisha's urban population lives in slums.[7] Bhubaneswar, the largest city and the state capital, has almost 40 per cent of its population living in slums. The bigger the city, the larger the slum population. The common perception is that slums are encroachments and that slum dwellers are lawbreakers who deserve to be treated with contempt.

Officials governing the cities are no exception. Hence, they tend to overlook slums when it comes to providing basic urban services such as drinking water, street lights, roads,

---

[7]Planning and Convergence Department, Government of Odisha, Odisha Economic Survey 2019-2020, February 2020, 115–16, https://pc.odisha.gov.in/sites/default/files/2020-03/Economic_Survey_2019-20.pdf. Accessed on 20 December 2023.

The Odisha Economic Survey 2019–20 has pointed out that the total population living in slums across the state was estimated to be 3.72 per cent of the state's population, which has been worked out to be 1.56 million people, the lowest in the country.

drainage and sanitation. Thus, these urban settlements are forced to perpetually exist in a precarious state devoid and deprived of basic services essential for decent living.

To understand the finer points regarding the genesis of slums in any city, one must examine the contributions of slum dwellers in making the city. When a city grows, large-scale construction activities, such as road laying, construction of housing and other commercial buildings, take place, which need a large number of labourers. No city can meet the demand for a labour force created by exponential construction activities with only its local labourers.

In such situations, labourers from nearby villages and other parts of the state, who are struggling to make a living in their places of residence, get attracted by the opportunities available in cities and readily migrate.

Gainful employment acts as a pull factor for this labour force, whereas a lack of adequate employment opportunities and facilities in rural areas act as push factors. Several acts and rules are made to ensure the welfare of labourers.

For instance, Section 34 of the Buildings and Other Construction Workers (Regulation of Employment and Conditions of Service) Act, 1996, mandates the provision of accommodation facilities (construction workers' dormitories, hostels or barracks with all essential infrastructure and amenities) within or in the proximity of project sites.

However, even large government projects fail to comply with such conditions. A lack of on-site accommodation facilities and a good and affordable public transport system leaves these migrant labourers with no option but to make their own arrangements and settle close to their work area by constructing a hut in a nearby slum. If they can't chance upon a slum in the vicinity of the site, these migrant labourers group together and establish a new settlement.

It is very common to find new slums coming up near

most big-ticket projects, and even after the completion of these construction projects, the newly established settlements remain and grow to become slum.

Generally, slums or low-income settlements are welcoming of new occupants to their settlement. Upon securing employment, the migrants approach the community leaders in the slum near the worksite (or the informal landlord, depending on the scale of the slum) and start residing in the new space with their permission.

Thus, the growth in size of a slum or low-income settlement depends upon the number and size of infrastructure projects taken up in the vicinity. No city provides sustainable housing infrastructure facilities to incoming migrants. Besides playing a critical role in constructing cities' infrastructure, slum dwellers also play a major part in the day-to-day maintenance of the city from driving cars, autos and cycle rickshaws to providing domestic help.

Cities also have sewer lines, a facility entirely managed by these slum dwellers, which, if not properly maintained, may result in serious health hazards. Slum dwellers have a resilient presence in the cityscape. They not only create the city but also run it. It will not be an exaggeration to say that cities will come to a grinding halt without slum dwellers.

Unfortunately, they have no place or proper house of their own in the city that they helped create. When it comes to providing basic urban services, such as water supply, these slums do not seem to exist in our municipal records. The absence of basic municipal services indicates that their place remains physically untouched by the city's governance.

In Odisha, during the implementation of AMRUT, I found the mapping exercise to be very interesting and enlightening as it helped unearth many hidden areas in cities, which were mostly slums that were left without piped water supply and were undocumented before the exercise.

I often faced the standard rhetoric from city officials and engineers that almost all areas of the cities had been covered with piped water supply. However, the ground reality was that the slums were completely ignored.

Taking cognizance of the apathy of the officials, I decided to mandatorily involve municipal councillors in the survey process. Because councillors' stakes and accountability are much higher than those of the officials, this strategy worked well and all the slums that had been either entirely or partially ignored till then were included in our action plan.

This strategy was further improved, and the PHEO/Water Corporation of Odisha[8] (WATCO) was made to mandatorily place the final action plan list of projects before the municipal council in a meeting convened exclusively for this purpose.

At the meeting, the executive engineers were asked to present the action plans to the councillors and explain how these projects would ensure 100 per cent coverage without leaving out any lane, by-lane or street in any ward. I'm happy to note that this new strategy ensured participatory planning involving public representatives at the grassroots level and yielded very good results.

Our zeal to use AMRUT for the social benefit of the state saw us taking one more bold decision, whereby the state provided two taps, one in the kitchen and the other in the toilet, to every urban poor family free of cost. The struggle related to sourcing and managing the water requirements of an Indian family disproportionately impacts the women of the household, who spend hours fetching water to meet their family's needs. This adversity increases with the poverty level.

After the implementation of AMRUT and with access to drinking water from the taps in their houses, women can

---

[8]WATCO is an ISO-certified, Government of Odisha-owned, not-for-profit company under H&UDD.

now engage in economic activities and contribute to the household income. Providing water through taps in toilets has greatly helped the cities achieve an open defecation-free (ODF) status, which further leads to higher standards of health and hygiene.

This, to me, is one of the biggest game changers in citywide inclusive sanitation (CWIS), which is being widely talked about today. The decision to provide taps to households was taken by the technical committee headed by me.

In hindsight I realize that it was a very big decision that generally officers hesitate to take, but somehow I took that decision and went ahead and implemented it across the state under AMRUT. Having worked with CM Naveen Patnaik as a collector in his district and later as his PS, his pro-poor governance ethos somehow rubbed off on me, which gave me the courage to make such decisions in the public interest. This experience of pioneering an extremely impactful initiative gave us confidence and tremendous satisfaction.

We soon started witnessing a transformation in the drinking water scenario in the nine AMRUT cities. The question we then faced was, 'With AMRUT, we could provide piped drinking water to almost all the houses in the nine AMRUT towns; it's a dream come true. What about the remaining 105 cities in the state? Are we justified in discriminating against people living in non-AMRUT towns just because their city has a population of less than 1 lakh (AMRUT covers cities only with a population of more than 1 lakh)?'

This question was nagging my conscience and disturbing my peace as I recollected the CM's words when I presented the water supply woes, 'Mr Mathi, we must do something.' Armed with the experiences and learnings gained from implementing AMRUT in the nine cities, I started visualizing universal coverage of piped water supply in all the 105 non-

AMRUT cities of Odisha irrespective of population. However, the following questions came before us:

- How do we conceptualize the new scheme?
- How to mobilize funds for the scheme?
- What are the sources available for tapping?
- How long would it take to complete these projects?
- Is my team capable of executing so many projects simultaneously across the state?

We started designing a state-funded scheme in urban Odisha along the lines of AMRUT; it focused only on the piped water supply component. The idea of a state-level scheme was born, which subsequently culminated in the BASUDHA[9] scheme.

The inequities in water supply were extremely disquieting for me because of the grave injustice meted out to the urban poor. The scope for attracting government funding is relatively higher for large cities like Bhubaneswar than for small towns because of the strong voices of the powerful lobbies in bigger cities.

At that time, most cities were encountering erratic and inadequate supply with low water pressure, which often led to conflicts between the public and administration. Water crises in cities were annual occurrences, especially during the summer, and the administration had a hard time dealing with public agitation, assembly questions, adjournment motions, court cases, adverse media coverage, etc.

Other issues, such as water contamination, drinking water mixing with sewage water and the spread of waterborne diseases with occasional fatalities, added to the problems of the already stressed and defamed PHEO. Many times,

---

[9]The BASUDHA scheme, launched on 23 November 2018, ensures universal access to potable piped water supply to households in rural and urban areas.

these incidents created discord between the state's Health and Family Welfare (H&FW) Department and the H&UDD, with each accusing the other for causing the crisis.

As the way forward, we decided to transform the existing system of inequity in accessibility and availability of drinking water into universal coverage of pipe water supply through BASUDHA. It is important to understand the foundations laid by us from 2017 to 2020 to understand and appreciate our journey leading to Odisha's Drink from Tap (DFT) mission in Puri in 2021.

The planning exercise started with habitation mapping, infrastructure assessment and gap analysis in early 2017. The rough cost estimation of the projects was done using a rule of thumb consistent with the processes and systems followed during AMRUT implementation.

The entire PHEO machinery was made to work on the in-house preparation of the project reports. In six months, the proposal for 555 piped water supply projects in 105 cities with the funds required (₹3,800 crore) was finalized. Interestingly, the department had thus far been receiving an annual budget of about ₹200 crore only for water supply, of which the PHEO had managed to spend about 90 per cent. Under the circumstances, our proposal for executing 555 projects with an investment of ₹3,800 crore raised two major concerns:

- Will the finance department approve a budget of ₹3,800 crore, a huge increase from the existing annual budget of ₹200 crore?
- Does the PHEO have the capacity to execute work worth ₹3,800 crore when the current spending capacity is just ₹200 crore annually?

Along with the AMRUT mission, the department was also implementing the Smart Cities Mission (SCM) in two cities of Odisha, namely, Bhubaneswar and Rourkela.

The SCM has two components: area-based development (ABD) and pan-city development. Of the investment under SCM, approximately 70 per cent is for ABD and 30 per cent is for pan-city initiatives. In Bhubaneswar, for the ABD area of 1,000 acres with a population of about 40,000, the investment proposal was for about ₹3,000 crore.

The per capita investment for ABD works out to be a whopping ₹7 lakh for approximately five years. In contrast, the per capita, one-time capital investment of ₹7,000 would benefit 35 lakh people through the universal coverage of piped water supply in the remaining 105 cities.

When ₹7 lakh per capita was proposed to be spent for creating urban infrastructures with smart city solutions, such as smart roads and Wi-Fi hotspots under the SCM, investing ₹7,000 per capita for providing safe drinking water to every home was considered highly justifiable.

To me, it appeared grossly unfair to invest ₹7 lakh per capita for improving the lifestyle of a small section of people when by providing ₹7,000 per capita, one of the basic necessities of life—water—could be provided to a larger population.

This comparison brought the irrational and skewed approach adopted in public investments to the fore. This fact was very unsettling and hurt our collective conscience. We also realized that the investment in the water supply programme could be staggered as the project was to be taken up over a three-year period at about ₹800 crore per year.

Though the scheme had a laudable objective, the question as to whether the PHEO would have the capacity to deliver such a mammoth task within three years was bugging me, and I was trying to fathom the actual time that the PHEO would require to execute the project.

Several other apprehensions were also buzzing in my

mind at that time, but providing drinking water to every home in urban Odisha was the most powerful driving factor for me to overcome them and go ahead with the proposals.

The tremendous satisfaction that I derived in implementing water supply projects in AMRUT cities, especially covering all slum areas, was a strong motivation for me to think big and go for universal water supply coverage in urban Odisha.

Throughout the journey, the only words ringing in my mind were, 'Mr Mathi, we must do something.' I would cease to unduly worry about overcoming the challenges because of the unwavering determination at the CM's level. I knew that with the strong conviction of Naveen Patnaik and his then PS Mr V.K. Pandian, they would find ways to secure funds for such a noble cause.

I had the conviction that the annual requirement of ₹800 crore for three years could very well be leveraged from various schemes and sources through convergence. I strongly believe that our CM would be immensely happy if we could realize this dream. For me, this was a perfect moment and a golden opportunity to chase my long-cherished dream. I did not want to miss out on the immense fulfilment that the scheme would bring.

THE ESSENCE

The test of our progress is not whether we add more to the abundance of those who have much; it is whether we provide enough for those who have too little.

—FRANKLIN D. ROOSEVELT

Slums are often perceived as encroachments in our cities, with the obvious suggestion of eviction as the solution. Ignoring their critical role in the creation and functioning of cities, they are often viewed as trespassers rather than enablers.

However, a deeper understanding reveals that slums are integral to the functioning of cities, and their existence is a result of our mismanagement and failure to provide basic amenities, including housing, to migrant workers. Notably, about one-fourth of the population in large cities resides in slums, and comprehensively addressing their needs is essential for equitable and inclusive urban service delivery.

Naveen Patnaik, the CM of Odisha, recognized the critical role of slum dwellers in cities and came up with the Jaga Mission to provide land rights and liveable habitats. The Jaga Mission garnered international recognition as the world's largest slum land rights and upgradation programme with several global awards.

Implementing the AMRUT scheme and designing the BASUDHA and Jaga Mission projects in Odisha, I learned valuable lessons about the significance of including slums in urban planning, the need for pro-poor governance and the need for the equitable distribution of public investments.

Providing two taps free of cost to every urban poor family and liberating the women of the slum households from the arduous task of daily water collection proved to be a transformative decision. With this newfound freedom, they could engage in economic endeavours and contribute to the financial growth of their households, thereby shattering the shackles of poverty and embracing a life of empowerment and dignity.

Empower with inclusivity:
transforming lives and building a better future.

4

POLITICAL WILL: AN ESSENTIAL INGREDIENT IN THE RECIPE FOR TRANSFORMATION

Having decided to embark on this ambitious journey, it was time for us to secure and commit funds for the programme. By then, we had completed state-wide mapping and the preparation of the action plan with the bulk cost estimate. The project funding required consultations with the finance department for budgetary provisions.

I was wary of the idea of taking up the proposal with the finance department because, for proposals of such magnitudes, their standard response is a big *NO*. I didn't want to be discouraged and dissuaded by anyone from going ahead with the programme.

Another inherent risk of approaching senior officers with such requests is that once they reject it formally or informally, they do not like it being pursued further or at higher levels.

So instead, I took the proposal directly to Mr V.K. Pandian, who had always encouraged and welcomed such people-centric transformative initiatives.

I briefed him regarding the precarity of the prevailing state of urban water supply with more than 50 per cent of the state's urban areas not having a pipe network and more than two-thirds of urban homes going without piped water connections. I also told him that whatever little supply they got was too erratic, irregular and often contaminated.

Moreover, there existed a huge disparity among the cities that were closer to the capital and those located in western and southern Odisha. Even within the cities, while a few areas were networked and getting excess water, most areas were deprived of piped networks, where people were managing with hand pumps and tube wells.

Mr Pandian immediately understood the situation and realized that this did not sit well with the image of the CM who had been at the helm of affairs for over 17 years by then. Having listened to me patiently, he said that this was a top priority for the CM and that it was his desire to see every home in Odisha have a piped water supply.

Given the scale of the problems and considering the impact that could be achieved, I told him that the required investment was not disproportionately high but reasonable. He looked me in the eye and firmly said that we should go ahead with the programme come what may.

The fund requirements to the tune of ₹3,800 crore were discussed in detail. I indicated that the expected additional budget from the state would be around ₹500 crore annually for the next three to four years over and above the present allocation of ₹200 crore per annum; the rest could be leveraged through the convergence of various ongoing schemes.

I was very positive about the department's capability to execute such a massive programme across the state. Furthermore, I had a belief that the learnings and hands-on experience from AMRUT would help us implement the programme in a timely manner.

I confidently committed to ensuring the completion of the programme within the next two to three years and before the general elections of 2019. I had a belief that the deep conviction and passion Mr Pandian saw in me and the trust he had in me convinced him to give me the go-ahead signal and to get the CM's approval, which he did.

After getting his approval, the programme got all the required formal sanctions from the finance department, planning and convergence department and chief secretary smoothly. Nevertheless, questions regarding the PHEO's capacity and my ability to complete the project within the stipulated timeline were raised, to which I gave a confident and affirmative reply that I would deliver as committed.

After the initial approval of the proposal, the details of the fund requirements and sources were worked out. The required fund of ₹3,800 crore was decided to be generated through a convergence of various ongoing schemes and sources, such as AMRUT, the District Mineral Fund (DMF), Urban Infrastructure Development Scheme for Small and Medium Towns (UIDSSMT), Odisha Mineral Bearing Areas Development Corporation (OMBADC), the Members of Parliament Local Area Development Division (MPLAD), Member of Legislative Assembly Local Area Development Scheme (MLALAD) and corporate social responsibility (CSR). A part of the funding was provided by the state's exchequer as well. Thus, the issue of mobilizing funds for the programme, which had seemed insurmountable at the beginning, was successfully managed.

POLICY REWRITTEN AND POLICY IN ACTION

I am obligated to give huge credit to my predecessor, Mr Injeti Srinivas, IAS, for formulating the Odisha Urban Water Policy 2013 when none existed. He conducted a situation analysis of the state-wide infrastructure and designed a 12-year improvement programme, which was split into three phases of four years each, following a cautious approach considering fund and capacity constraints.

AMRUT and the SCM were implemented thereafter, and the learnings that we gained from them brought a paradigm shift in our approach towards addressing the most challenging task: covering all lanes and by-lanes of every city with a pipe network along with providing tap water supply to every home in urban Odisha.

We realized that achieving these by 2027, as envisaged in the 2013 Odisha Urban Water Policy, would be too long and the state could not and should not delay achieving universal coverage by another 12 long years. Therefore, we decided to achieve the targets by 2020, advancing the timeline by seven years. Thus, the policy was rewritten, not on paper but in action.

It was an inspiring and ecstatic moment for me and my team when the CM of Odisha launched the BASUDHA scheme at the Cuttack Indoor Stadium on 20 October 2017. I looked at the CM and Mr Pandian with gratitude, and the latter responded with a thumbs up. Seeing immense happiness on their faces, I knew that failure wasn't an option for us on this journey. With the launch of BASUDHA and the need to quickly roll out all 555 projects, we started working on team building and innovative strategies for achieving the seemingly unachievable.

THE ESSENCE

Political will is the decisive force that turns intentions into actions, paving the way for progress and shaping the course of nations.

—KOFI ANNAN

When Naveen Patnaik had said, 'Mr Mathi, we must do something,' after I had given him an overview of the drinking water scenario in Odisha, it was an expression of political will and a confidence-boosting go-ahead. The experience leading to the launch of BASUDHA in 2017 taught me that transformative initiatives that demand huge financial outlay should first be taken up with the top political boss before going through the regular bureaucratic channel.

Conventional bureaucracy has a tendency to kill such high-stakes initiatives in the budding stage without appreciating their social impacts. So, I took the matter directly to the CM through his PS, and the strong political will for the initiative helped me overcome all the hurdles.

To leverage political will, it is crucial to present the leaders with the vision, outcomes and social impacts that the initiative can generate with speed and scale being the critical factors.

Political will can provide the impetus needed to overcome bureaucratic hurdles and mobilize resources for transformative initiatives. It creates a ripple effect among different stakeholders, encouraging them to collaborate and work towards the common goal. Additionally, it helps garner public support.

The Odisha urban water programme is a classic example of how strong political will could bring about transformation in the urban drinking water sector.

From words to action:
strong political will paved the way for
clean water in urban Odisha.

5

PROCESS RE-ENGINEERING

One of the most important lessons learnt from AMRUT's implementation was that the appropriate sizing of projects is crucial to ensure successful and effective execution. We realized that it is prudent to go for small- and medium-sized projects instead of large contracts of higher value for the following reasons:

- Local contractors are ineligible to participate in tenders of high-value projects, and the contracts are given to bigger players mostly from outside the state.

  Besides limiting competition (sometimes resulting in single bids, which necessitates retendering and wastes time), high-value tenders are mostly taken at a high premium, thereby defeating the very purpose of tendering (because most big players are not domiciled in the project area, and they tend to engage small local contractors, who eventually deliver the work and add an additional margin to the price bid).
- If disputes arise in the execution of larger projects, the litigation process stalls the execution of the entire project till the disposal of the litigation. If the full project is broken down into smaller independent projects, disputes arising in one will not affect the execution of the rest.

Keeping these facts in mind, we went for small- and medium-sized projects that can be tendered and implemented locally at the executive engineer (EE) division level, mostly with the local contractors with an implementation period of 6–12 months.

This approach we adopted had emerged as an outcome of our learning from the implementation of AMRUT, though it was quite opposite to the advisory of the Government of India, issued under AMRUT.

To decide the size of the contract and the number of projects that can be implemented division-wise, we had to assess the number of local contractors available in each division. From the assessment, we realized that many divisions had only a few contractors with limited financial capacities.

Such situations can lead to tender cartelization. Realizing the impracticality of implementing many projects with these few local contractors, we felt the need to look for additional contractors who were engaged by other sectors and departments, such as roads and water resources, across the state.

A consultative workshop was organized with the contractors' associations and their members to seek suggestions for bid criteria that would enable their participation in the tenders. Thereafter, the SBD was suitably amended to facilitate the participation of contractors from other sectors.

This resulted in the onboarding of many new contractors for water supply projects across the state, thus successfully avoiding cartelization and facilitating increased competition and faster implementation.

To provide a piped water supply connection to every home within three years, we decided to go for a massive piped network infrastructure for the distribution. Regarding the supply, wherever it was anticipated that drawing water from far-off rivers would require a much longer execution time, we opted for interim solutions such as water supply production wells. Further, rationing of the existing supply, with the addition of new areas and an increase in the pipe network, resulted in equitable distribution across the city. Many contractors were engaged to ensure that the piped water supply network projects were completed in six to nine months.

After addressing the contractor issue, the next challenge to be resolved was the in-house capacity constraints at the departmental engineer's level. Traditionally, detailed project reports (DPRs) are prepared for infrastructure projects by engaging consulting agencies due to a lack of in-house capacities.

This practice had several disadvantages, such as the availability of limited consulting agencies, longer time for DPR preparation, consulting agencies resorting to desk-based DPR without the necessary field verification and the preparation of unrealistic estimates.

Additionally, due to a lack of in-house capacities, the DPRs prepared by external agencies get approved without due scrutiny by the PHEO. Adopting such DPRs has often resulted in tender premiums being as high as 80–90 per cent in some cases, which forces the department to bring the matter to the finance department for concurrence, thereby causing considerable delays. This was the scenario when the PHEO was executing fewer projects at the divisional level. Upon launching BASUDHA, we had to develop a new strategy to address these issues.

During one of my personal visits to Chennai, my home town, I met Mr Vijaya Rajkumar, the then managing director (MD) of the Tamil Nadu Water Supply and Drainage (TWAD) Board to understand how Tamil Nadu was able to implement a large number of piped water projects. The TWAD Board had executed numerous pipe water supply projects long ago, even in the small villages of Tamil Nadu; my village, comprising 150 houses, had piped water supply as early as 1990 itself. I understood that they were using some software to prepare DPRs, which made the DPR preparation process easy, accurate and fast.

We decided to adopt the same approach and requested the TWAD Board MD to help train our engineers. A team of around 30 engineers were deputed to the TWAD Board to undergo training. Besides training our team for a week, the TWAD Board also shared their software free of cost. The trained team returned and trained other engineers in the PHEO divisions.

Subsequently, our engineers started preparing DPRs better and faster than the consulting agencies. This capacity-building effort led to the in-house preparation of all DPRs without having to engage consulting agencies, hence saving time and money. The familiarity of our engineers with field conditions enhanced the accuracy of the DPRs and resulted in realistic estimates.

This led to getting more competitive bids and a reasonable tender premium in the range of ±3 per cent. These measures not only saved time and money but also spared us from obtaining the finance department's approval for every tender.

Another important strategy that helped implement the project faster was reversing the embargo on the departmental procurement of pipes imposed by the government a decade ago. This ban was imposed to curtail scams involving the unnecessary procurement of pipes at the end of the financial year. These pipes remained stacked in piles and stocked in stores without being used in a water supply facility; they were merely purchased to utilize budget allocations and avoid the surrender of funds.

In a distribution network project, the material component is about 70 per cent, with pipes and fittings as major constituents, and the labour component is about 30 per cent. The contractor is compelled to cover 70 per cent of the cost for pipe procurement, which must be done in advance, and it takes longer for them to get paid for materials after the completion of the work. We realized that if material procurement was removed from the scope of the contract, then the cost of the contract would reduce substantially, thus allowing more contractors to participate in the bidding process.

This would also facilitate the easy mobilization of the workforce and resources for the contractor, leading to faster implementation. Reducing the contract value would also enable local contractors to bid for multiple projects, leading to more competition and a lesser tender premium.

We were fully convinced that the departmental procurement of pipes would be the right strategy for the faster implementation of a large number of projects and decided to request the government to lift the ban on the departmental procurement of pipes.

Some of my well-meaning colleagues advised me against this idea because they were apprehensive of a negative response from the finance department and a possible misunderstanding of my intentions. I pondered over these concerns for quite some time and came to the conclusion that there was no other option than to go for departmental procurement to speed up the implementation process.

My bold decision made many heads turn in disbelief. The key decision-makers in the finance department were sceptical of my proposal despite my earnest pleas that it was absolutely necessary to enable the state to complete the projects in time using AMRUT funds.

All my earnest efforts to make them see reason went unheeded. I submitted a formal proposal seeking an exception for the departmental procurement of pipes and clearly highlighting the likely consequences if my proposal was not accepted. As expected, the finance department promptly rejected the idea.

The passion that was driving us to realize the dream of providing safe drinking water to every home did not allow me to accept the setback and shelve the proposal. I approached senior officials and fervently pleaded my case knowing fully well that I stood the risk of my intentions being suspected because the finance department's apprehensions regarding the proposal were well founded.

Seeing my strong conviction, one of my well-meaning advisers, Mr R. Balakrishnan, IAS, who was the development commissioner at that time, suggested raising the issue before the apex committee. This committee was chaired by the CS and included the development commissioner and additional chief secretary to the CM as members.

With a glimmer of hope, I met with the finance department officials who had rejected our proposal to understand all their apprehensions regarding allowing the departmental procurement of pipes.

Thereafter, putting myself in their shoes, I discussed their apprehensions with my team and came up with comprehensive strategies to address each of their questions.

Now strengthened with answers to the finance department's concerns and to ensure a foolproof system for preventing the possible misuse of funds, I sent a note to the CS requesting him to consider placing my proposal before the apex committee.

At the apex committee meeting, I was given a chance to present my case. Before the secretary of the finance department could voice his concerns, I made a strong case by listing all their possible apprehensions regarding the proposal and the strategies proposed by us to address them with checks and balances.

With confidence and conviction, I proposed the following to the apex committee as part of our strategy:

- Pipe procurement will not be commenced unless the water source is confirmed and the network project is successfully tendered and kept ready for award.
- Hundred per cent third-party verification of the pipes procured and laid will be conducted.
- The payment for pipe procurement and their laying will be released only after the receipt of the third-party verification report.
- Reconciliation of the quantum of pipes procured and laid will be undertaken annually.
- Procurement will be decentralized at the divisional level and limited to ₹10 crore (₹100 million) per tender.

To avoid possible hesitation in taking a permanent decision for such a major issue, I suggested that special permission may be granted for two years only for this programme, which could be reviewed thereafter.

I concluded my submission with a polite but strong note that if the requested special permission was not granted, AMRUT implementation would be seriously hampered. Furthermore, the state may be penalized with substantial cuts in central grants due to the delayed implementation of the scheme.

After arguing my case before the apex committee, I realized that I had put my personal credibility at stake by pushing this agenda too far. However, I was also satisfied that I had done my best to pursue the cause, which I believed was far more important.

Finally, the proposal was approved, all the safeguards we had suggested were accepted and the ban on pipe procurement was lifted for one year for AMRUT projects.

With this success, we marched ahead with a smile on our faces.

At the end of one year of execution, we fulfilled our commitment by saving ₹27 crore for the exchequer and achieved an accelerated pace of programme implementation. I went back to the apex committee with a DPR that highlighted the total quantity of pipes procured and laid and the amount saved. I thanked the committee and the government for the landmark decision to permit the departmental procurement of pipes.

Citing the successful implementation of AMRUT projects after pipe procurement was allowed, we requested the government to extend similar permission for BASUDHA projects as well and extend the period for AMRUT projects by one more year.

The government saw sense in our proposals and promptly approved our request. After carrying out the departmental procurement of pipes with the necessary checks and balances, we concluded that it played a crucial role in saving time and money apart from resulting in other advantages such as existing local contractors bidding for multiple projects leading to more competition, lesser premium, faster implementation and the prevention of tender cartelization.

Realizing the benefits of departmental pipe procurement and seeing the credibility of the urban team in sincerely implementing the water supply agenda, the finance department later delegated their power of approving departmental pipe procurement to the secretary of H&UDD. Now, departmental pipe procurement has become the norm for urban areas but with safeguards.

Thus, we cut down the time delays that could happen in procurement. Then, we started identifying process delays and ways to fix them for barrier-free and speedy project implementation. Giving separate administrative approval for each project was one such major source of delay. Each project required administrative and technical approval of the DPR before it was tendered to select an executor.

After the approval of the action plan with the list of projects at the government level, the DPRs of individual projects undergo scrutiny for administrative approval, which is a bureaucratic decision-making process involving two aspects: commitment of funds and approval to execute the project.

Any water supply project valued up to ₹15 crore is approved at the level of the minister-in-charge of the department. Projects valued at more than ₹15 crore require the finance department's acceptance before they are approved by the minister. This process of obtaining administrative approval takes several months and involves three offices of the government, namely, the engineer-in-chief (EIC), the H&UDD and the finance department.

Administrative approval for 555 projects in this manner would have involved a lengthy bureaucratic process of individually dealing with 555 files resulting in abnormal delays even before the commencement of the tendering process. Therefore, we decided to change the usual process of floating individual files for administrative approval and go for composite administrative approval at the minister's level with the finance department's concurrence in a single file by clubbing the 555 projects together.

Because every case record requires administrative approval of that project as an essential document to stand the scrutiny of audit, the engineers' apprehension was addressed through formal instruction to all field offices that one certified copy of the composite approval could be kept in every project case record to meet the documentation protocol. This file process saved us from unnecessary paperwork and consequent delays.

Initially, the BASUDHA mission was launched with 555 projects after getting a certification that all habitations had been covered and no area had been left out. With the launching of the mission by the CM and given the publicity that the event created, information regarding areas left out from the project list started reaching us from various quarters.

We realized that the engineering cadre's strength and capacities had not kept up with the increasing scope of work and responsibilities. In many small and medium cities, there were only one or two engineers with pump operators and linemen who delegated their responsibilities to informal and invisible workers. The engineers were heavily dependent on such weak field machinery for all field operations.

To our dismay, we found that field information for the projects had flowed into the state level from these invisible and informal workers and that the state authority had collated and presented the information as received and without any further validations. The PHEO was not connected to the people on the field. Only the reported problems were addressed, leading to public scepticism that only the influential and elected representatives get their issues attended to by the PHEO.

Up until that point, the general approach of the PHEO had been to manage water supply affairs smoothly, primarily by maintaining the status quo, to avoid any major crises and secure their jobs. There was no connection with the consumer. This pattern of governance in urban centres forced people to approach elected representatives either directly or through middlemen for the resolution of their problems. Because of this disconnect in the system, many areas were left out during the initial survey and mapping exercise.

With the launch of BASUDHA, extensive information, education and communication (IEC) activities were taken up, and the message carrying the intent of the government to provide piped water supply to every home in the cities reached the grassroots; thus, we started getting information about the areas left out from the project list. I must admit that it took some time for the IEC to unfold and manifest with the addition of new projects from time to time.

With hidden, neglected and left-out areas getting added, the number of projects swelled from 555 to more than 980. Though it was a frustrating experience to receive new project proposals time and again, knowing the disconnect and inefficiency that prevailed in the system, I accepted each one of them to honour our commitment to ensuring that no house was left out from the pipe water supply system.

With these interventions, we managed to overcome major issues starting from scheme formulation to the tendering process. Once tendering was completed and project execution commenced, several challenges in field implementation started unfolding and our journey continued gaining momentum.

THE ESSENCE

The art of progress is to preserve order amid change and to preserve change amid order.

—ALFRED NORTH WHITEHEAD

The implementation of AMRUT inspired us with many insights, especially on what not to do in Odisha's context. One of the key takeaways was that while limited contractors were available to execute large projects, many contractors were available to execute small-scale projects. Hence, there was a need to split up the entire water supply project into small projects based on component-wise expertise, such as the construction of intake wells, pump houses, WTPs, ESRs, underground storage reservoirs and laying of transmission and distribution pipelines, to tender them as separate projects so that the projects became more competitive and all components got executed simultaneously.

The approach of going for smaller component-wise projects enabled local contractors to participate, increased competition, reduced litigation risk and resulted in speedy execution. Appropriate project sizing is key to more competitive bidding and faster implementation.

Along with the tender reforms, process re-engineering is vital for drastically cutting down bureaucratic delays. Composite or clubbed administrative approval was done for all 555 projects in one go instead of separate approvals for individual files.

The policy for pipe procurement was also reversed to allow the department to procure pipes, which increased competition due to reduced project costs and saved time and money.

The capacity building of in-house engineers to prepare DPRs resulted in the preparation of realistic estimates and consequently reduced tender premium to within ±10 per cent. This eliminated the need for retendering or seeking the finance department's approval and consequent delays.

Adaptation, capacity building and process re-engineering are the keys to successful project implementation.

6

HONOURING COMMITMENTS

FINISHING THE UNFINISHED

When we started drawing the contours of AMRUT 1.0's implementation, I reviewed what had been done in the past and where we were at that time. I noticed that the asset creation pattern was very strange. Though the requirements for new projects for the underserviced areas were huge, annual budget allocation was disproportionately less due to the PHEO's capacity constraints, which resulted in the low utilization of allocated funds. Hence, the annual budget allocation was limited to just 10 per cent more than the previous year irrespective of growing needs.

There was a tendency to start new projects every year based on political/administrative compulsions or requests rather than providing funds to complete the ongoing projects. Eventually, many incomplete projects piled up. Several small projects, which otherwise could have been completed in six months to a year, took more than five to six years to complete. Many incomplete projects were almost forgotten for a long time.

Though the state had been making significant investments every year, we were not reaping the benefits of the investments because the projects were not getting completed. The focus was lacking, and priorities kept changing every year. We understood the situation and prioritized completing the unfinished projects.

All divisions were instructed to take stock of all the incomplete projects, work out the fund requirements and submit proposals. An advisory was also issued that unless all remaining incomplete projects were completed, no new projects could be started by any division.

Using this new strategy, we could complete a significantly large number of incomplete projects in record time, which had been stalled for a long time. This had a huge impact on the ground and pushed the PHEO towards outcome-oriented project implementation and budget utilization. In the process, the PHEO was made to redefine its focus towards outcomes instead of simply utilizing the allocated budget.

Earlier, the completed projects were usually inaugurated by a local Member of the Legislative Assembly (MLA) or other political dignitaries as these projects were of little value and generally delayed. To give momentum and importance to the drive to complete unfinished projects, it was announced that all newly completed projects would be clubbed together and inaugurated by the CM at a state-level event only.

This strategy created the desired momentum and energy across all PHEO divisions throughout the state leading to an expedited completion process. Bringing in the CM in this process also prevented local pressure from political functionaries from starting new projects and leaving aside unfinished ones. This also ensured that the funds were exclusively used for completing the projects and not diverted elsewhere.

THIRD-PARTY VERIFICATION

One of the main reasons for the government to ban departmental procurement of pipes was that in several instances in the past, pipes were procured in large quantities and stacked in the field but not laid. Several cases of fictitious procurements and laying of pipes were unearthed during field verifications for projects that existed only on paper. I risked my personal credibility while seeking and securing approval for the departmental procurement of pipes for the speedy implementation of water supply projects.

This was a special permission given only to our department on my assurance to the government that there would be appropriate safeguards and checks in place to ensure genuine procurements. With a very large number of projects to be executed (more than 900) and with each project having multiple work fronts involving innumerable contractors and field functionaries, we realized that the PHEO did not have the adequate capacity to closely monitor field execution to ensure integrity in the execution process.

Because the scale and size of the programme were so overwhelming compared to the in-house capacity to effectively supervise the implementation of the projects, a third-party monitoring agency (TPMA) was engaged to carry out the field verification process alongside the pipe laying work. The fictitious procurement of pipes was a major apprehension, so we mandated physical verification of the pipes stacked in the field with reference to the procurement specifications before they were laid to ensure adherence to the prescribed quality standards.

As we had assured the government of ensuring integrity in the implementation process and as my personal credibility was at stake, I preferred 100 per cent verification of the pipes procured instead of the usual practice of random sample verification.

To ensure a fair and independent verification and certification process, it was decided that the TPMA would be engaged and paid by another agency of the H&UDD

and not by the implementing agency—the PHEO. The State Urban Development Agency (SUDA) was entrusted with the responsibility, and earmarked funds were provided to SUDA to meet the cost of the third-party verification process. Field verification and certification by the TPMA were made mandatory prerequisites to pass running and final bills.

A separate dedicated WhatsApp group was created at the state level to facilitate the free flow of information and barrier-free communication among the TPMA, all divisions and the state authorities. Opting for third-party verification paid off with the achievement of the desired quality and quantity of materials procured and laid. This was a smart move by us to demonstrate the sincerity of the department in ensuring fairness in the procurement and implementation process.

But we didn't stop there. To double-check that no compromise was made in the fairness and accuracy of field verification by the third party, another layer of final field verification was established through a multimember committee. Thereafter, the task before us was to ascertain whether the project had resulted in the achievement of the intended objectives: availability of water in sufficient quantity and pressure, even at the farthest points of supply and in all the streets. A unique project outcome certification process was conceptualized and introduced.

INTERDEPARTMENTAL COORDINATION

Just when we were increasing the momentum and project implementation was in full swing, I received a frantic call from the EE of Bhubaneswar Division III. He informed me that the contractor engaged in laying the water supply pipeline in his division had been arrested and taken into police custody for alleged unauthorized road cutting.

Upon enquiry, we discovered that the EE of the Public

Works Department (PWD) had filed a First Information Report (FIR) charging the contractor with the unauthorized cutting of a PWD road.

This enraged me as I could not stand water supply contractors getting arrested for executing high-priority work duly awarded by the government. I had to intervene and confront the EE who filed the FIR and the inspector-in-charge of the police station who had registered it. I took up the issue with the PWD secretary and the police commissioner of Bhubaneswar to ensure the immediate withdrawal of the case and release of the contractor.

This sent a clear message to all the stakeholders that the provision of piped water to every home was a top priority for the government, which needed to be executed on a war footing. If road cutting was required, it would be done, and road restoration would also be ensured by the PHEO at its cost to the satisfaction of the road-owning agency. If any FIR was filed for doing this lawful duty, it had to be filed against me, the H&UDD secretary, and not against the contractor executing authorized work.

This strong response reverberated at the highest levels of the PWD and the police and ensured the sensitization of all their field officials. Such incidences were never repeated at project sites across the state. This incident also brought to the fore the need for coordination amongst all road-user agencies, which require road cutting for laying or operation and maintenance (O&M), to ensure that repeated road cuttings are minimized and avoided.

However, government departments always tend to operate in silos with little idea about what the other departments are doing in the field. Therefore, it is always imperative to start any project with team building at the field level by involving all the associated departments to ensure coordination and mutual support.

From my experience, I firmly believe that strong leadership, effective communication and understanding each other's priorities and constraints are crucial for building effective teams. Government functionaries working in different departments must understand the larger picture and work towards achieving the government's common goal rather than taking an individualistic and department-centric partisan view.

This also reiterates the need for government functionaries to have a collaborative approach towards delivering public services by involving multiple departments. One important learning from this experience was the need to set up a coordination mechanism at every city level, at least for big-ticket projects. This would help sensitize various functionaries of the relevant departments from the beginning and resolve interdepartmental issues from time to time for faster project implementation.

OUTCOME CERTIFICATION

The certification process was entrusted to a multimember committee comprising the district collector's representative, a PHEO official, a municipal official and elected and community representatives. Because the committee consisted of non-technical and non-official members, a simplified format with easily understandable and verifiable parameters was devised to capture the essential outcomes to be assessed for the project outcome certification.

This committee was mandated to visit the project areas to verify and certify the achievement of the intended project outcomes. It was stipulated that without the project outcome certificate, the case record could not be closed and final payment could not be made.

This strategy was meant to have another layer of verification involving multiple members, including local public and community representatives along with the officials, over and above the third party to verify the project outcomes.

This strategy ensured social auditing with the participation of the local community in the verification process. It also helped improve public perception regarding the fairness of the implementation process.

Furthermore, these multilayered verification processes involving third-party verification and social auditing instilled a sense of carefulness and accountability in the minds of the executing agencies. Owing to the strength of the robust strategies put in place to ensure project implementation in accordance with the guidelines issued, two cases of deviations made on the field were identified in time before any damage was done.

The lapses committed by two EEs in placing orders for pipe procurement were also found out in time, and the procurement orders placed by them were cancelled, thereby averting possible financial losses and violation of the conditions stipulated by the finance department while approving the special permission for the departmental procurement of pipes.

I took strong exception to this deviation and ordered swift and strict disciplinary action to suspend the two EEs. I also made a detailed report on the lapses made and corrective actions taken, including cancelling the procurement orders and taking disciplinary action against the erring engineers, and shared it with the finance department to maintain transparency and keep them updated on the developments.

The timely identification of lapses, prompt corrective action, strict disciplinary action against erring officials and wide publicity of the action taken helped send a strong and clear message to all field officials to refrain from indulging in violations. Thereafter, such violations seldom occurred.

THE ESSENCE

Integrity is doing the right thing,
even when no one is watching.
—CHARLES MARSHALL

Till completion, unfinished projects are non-productive assets that depreciate fast and are a waste of resources. By prioritizing the completion of these projects, we ensured that resources were best utilized without diverting focus to starting new projects.

With the special permission obtained for departmental pipe procurement, the onus was on us to ensure integrity in pipe procurement and laying. We enforced checks and safeguards, including 100 per cent verification of pipes procured through a TPMA, to verify the materials procured and laid. It worked!

We realized that people could mock our claims of project completion and the achievement of universal coverage unless a foolproof verification protocol involving independent stakeholders was established. By involving multiple stakeholders in the validation process, which includes field verification, such as local public representatives, community-based institutions, city and district administration representatives and civil society organizations, we ensured that the process was fair, transparent and trustworthy.

A blueprint for success:
putting the necessary checks and balances
in place with citizen monitoring measures.

7

POLICY, AN ENABLER

LAST-MILE CONNECTIVITY

Generally, in a piped water supply project, once the laying of the pipe network and other infrastructure construction is completed, the project is closed, final bills are settled with the contractors and engineers assume that their responsibilities are over. Connecting individual houses to the network is always perceived to be the responsibility of the house owners. When pipeline network projects are implemented, it is essential that HC work is simultaneously taken up.

This approach would help reap the benefits of the investment quickly and fully besides eliminating the need for repeated road digging and restorations, which further results in reduced inconvenience to the public and monetary savings in road cutting and re-laying work.

Though a large number of projects were completed, HCs did not increase proportionally. Without HCs, which is last-mile connectivity, the intended outcomes of the investments made could not be realized. Despite our best efforts to create local awareness amongst the public and put pressure on the PHEO machinery to ensure 100 per cent HCs, the pace of the process was not encouraging.

A deep dive into the factors responsible for the inadequate response for HCs revealed that the cumbersome application process was a major bottleneck, which involved submitting several documents (14, to be precise) with the application.

The requirement of numerous documents was the major deterrent for many from applying for an HC. One of the required documents was proof of ownership or lease or a rental agreement for the house. In Tier II and below cities, houses are generally rented out without a formal rental or lease agreement, whereas in bigger cities, though many leases are formalized with a proper rental/lease agreement, they are seldom registered.

In many cases, house owners reside elsewhere, which makes it difficult for tenants to obtain the owners' signatures on the applications. The plight of slum households was much worse because most of them were located on government land without any document to prove landownership; hence, they were rendered ineligible for water supply HCs. In 2017, Naveen Patnaik took a historic decision to recognize slum dwellers by granting them land rights.[10]

When others resorted to eviction to deal with slum issues, Naveen Patnaik saw slum dwellers as co-creators of the cities and decided to empower them with land rights and liveable habitats with all the necessary infrastructure and services. When I asked why ownership proof or rental/lease agreement was sought for providing HCs, I was told that the documents were required to avoid landownership-related litigations in courts in the future.

On being asked for details of litigations faced by them, they could not provide any information and I understood that it was unfounded fear. We decided that as an alternative

[10] The Odisha Liveable Habitat Mission (The Jaga Mission) was launched by the CM in May 2018. It aimed at transforming urban slums into liveable habitats with all the necessary civic infrastructure and services across the state and making Odisha the first state in India to be slum-free by the end of 2023.

to the landownership document, an indemnity undertaking may be obtained from the applicant, which would protect the PHEO and the government from any future litigations.

Other documents required, such as tax payment receipts, house plan approval and contractor licence for an HC, were equally difficult for the house owner to provide, thus making the whole process painful.

Looking at the number and nature of documents sought to sanction an HC, we realized that our approach and intent had been to discourage rather than encourage the people to obtain HCs. The need for the other documents sought was critically reviewed, and many were dispensed with.

With these revisions in the application format, the submission process was simplified and streamlined, which encouraged people to apply for HCs. I continue to insist that when roads are dug and pipelines are laid, work on HCs also needs to be undertaken simultaneously to enable road closure and restoration only once and to avoid repeated road diggings for new HCs.

This simultaneous process of providing HCs immediately after the street pipeline is laid is not possible unless applications for them are obtained from all houses on both sides of the street, and these are processed and sanctioned.

Subsequently, I found out that as many as 36 different documents were listed as mandatory documents to be submitted along with the application for piped water HC in various other states, which could be one reason why the country has achieved less than 50 per cent HCs in urban areas so far.

The Jaga Mission later received national and international recognition, including the UN-Habitat's World Habitat Award twice, once in 2019 and again in 2023.

I realized that doing away with several unnecessary documents, especially the land-related documents, and simplifying the process[11] had led to a massive jump in the number of HCs in the state from 3,00,000 in 2018 to over 9,00,000 in 2021. Additionally, we have not faced any land-related litigation in the courts so far, which was the PHEO's major apprehension; this proved that it was just another myth waiting to be broken.

EXEMPTION FOR ROAD-CUTTING CHARGES AND PERMISSION

After an HC was sanctioned by the PHEO, the applicant was required to approach the municipality and apply for road-cutting permission. The municipality prepared an estimate for road cutting and restoration and levied a charge that the applicant had to pay to obtain formal approval.

Thereafter, the applicant had to engage a plumbing contractor on their own and execute the work in coordination with the PHEO. This entire process was full of hassles and too arduous because the public had to rely on contractors and middlemen to manage the offices and processes and had to pay hefty charges for the service.

This state of affairs also encouraged those who tried to escape the hassles of getting formal connections to resort to obtaining illegal HCs with the help of field-level functionaries of the PHEO and municipality by greasing the palms of the abettors.

Our objective was to ensure hassle-free connections and prevent unauthorized connections, so we decided to put an end to the prevailing system. We decided to exempt the public from paying road-cutting charges and do away with

[11]The Odisha Gazette Notification No. 1670 dated 2 December 2015 (https://world-habitat.org/world-habitat-awards/winners-and-finalists/jaga-mission/).

the requirement of seeking road-cutting permission for HCs.[12]

Though this state action had the potential of being perceived as a move to usurp the authority of the local bodies in terms of levying road-cutting charges and granting permission, we went ahead with this reform measure by invoking the overarching powers of the state under the municipal acts in the larger public interest. The loss of revenue/compensation for road restoration has been made through the annual provision of adequate funds to urban local bodies (ULBs) for road construction and maintenance.

PAYMENT PROVISIONS FOR HOUSE CONNECTION CHARGES IN INSTALMENTS

The upfront connection fee of ₹3,060 was then mandatory to apply for an HC, which proved to be another stumbling block for many as they could not mobilize the amount immediately. We had undertaken mass HCs as a special drive for execution in tandem with pipeline laying to cut down costs of repeated road restorations and prolonged inconvenience caused to the public. We decided to exempt the urban poor from paying the upfront connection fee and allow others to pay in easy instalment options of ₹100 per month for three years.[13]

All these enabling policies facilitated the entire process from application to actual connection and made our journey towards achieving 100 per cent HCs faster.

---

[12]The Odisha Gazette Notification No. 5 dated 1 January 2019 (https://urban.odisha.gov.in/sites/default/files/2023-03/Drink-From-Tap-Mission.pdf).

[13]The Odisha Gazette Notification No. 5 dated 1 January 2019 (https://urban.odisha.gov.in/sites/default/files/2023-03/Drink-From-Tap-Mission.pdf).

COMMUNITY PARTNERSHIP

Providing HCs to all the houses in a locality simultaneously required large-scale mobilization of the households, which is possible only when the local community is involved.

Therefore, we formed water and sanitation (WATSAN) committees at the ward level to ensure the participatory involvement of local communities. The WATSAN committee was headed by the local councillor and had a few other prominent members, including the president and secretary of the local women's self-help groups (SHGs). One of the women SHG members was selected as the secretary of the WATSAN committee and designated as the water monitor of the ward.

The WATSAN committees were encouraged to organize Jal Jogan Melas, a citizen connect programme on water supply, with the support of the PHEO and municipal officials to create awareness about the water supply programme and facilitate the submission of applications for HCs. An incentive of ₹100 per connection was paid by the PHEO to water monitors to encourage them to mobilize the local public to apply for HCs.

THE ESSENCE

The right to water and sanitation is a human right. No one should be denied access to safe and clean water simply because of where they live or their income level.

—BAN KI-MOON

Policies, rules and regulations are created to work for the people and not act as stumbling blocks. Unfortunately, these remained archaic as we failed to make the necessary periodic amendments considering changing needs. The key is to keep reviewing and revising them to keep up with changing times. We rewrote the policies wherever and whenever required in line with people-centric priorities.

To overcome this challenge, the HC application process was streamlined, reducing the required documents from 14 to just 2 to remove the administrative barriers. Further, the responsibility for executing water supply connections was shifted from consumers to the government, and citizens were exempted from paying road-cutting charges and seeking road-cutting permissions to encourage authorized or lawful connections. The provision of easy instalment options for payment and fee exemptions for the urban poor helped simplify the entire process, thus accelerating the journey towards achieving 100 per cent HCs. WATSAN committees were formed at the ward level to facilitate HC applications and incentivize water monitors to mobilize the community.

From streamlining application processes to empowering slum dwellers, Odisha has ensured last-mile connectivity for piped water supply in every home irrespective of its location or the income level. We ensured that policies do not just remain on paper but bring a transformation in the lives of the people.

Policy must always work for the larger interest of people, not against it. Flexibility is the key.

8

JAL SATHIS: THE COMMUNITY PARTNERS

During project implementation, we began realizing that establishing community connect was vital to mobilize households for obtaining piped water supply connections as soon as the pipeline networks were laid. This also ensures that the local community supervises the execution of work. As we were progressing, we comprehended the need to regularly deepen our engagement with the community at the grassroots level instead of limited engagement with the WATSAN committee.

Though the WATSAN committee was intended to exercise overall supervision and monitoring in the field, the ward monitor could play a more proactive role beyond facilitating new connections and coordinating the activities of the WATSAN committee.

With the manyfold increase in household connections, the consumer base for the projects increased significantly. The functions and responsibilities of the PHEO and the expectations of the public from the PHEO also rose significantly. This warranted a corresponding increase in human resources through the enhancement of the cadre strength of engineers and supporting personnel.

Augmenting cadre strength would be a long-drawn process along with getting approvals at various levels, and completing the recruitment process thereafter may take several years to materialize. I nipped the idea of augmenting the cadre strength in the bud for two reasons: first, the lack of aptitude and interpersonal skills of many field officials in providing customer-friendly services, and second, the heavy recurring financial implications that an increase in cadre strength would entail.

The other alternative was to outsource work to private agencies through contracts. To me, both options were unattractive, time-consuming and less than effective. Instead, I strongly believed that enhancing the role of water monitors, who had already been chosen from the local communities and trained and who functioned as part of the grassroots-level teams, would give better results.

Thus, the idea of Jal Sathis was born as an improved and upgraded version of water monitors. Since then, Jal Sathis have filled the gap in the community connect and are acting as a bridge between the consumers and the PHEO.

With appropriate orientation and training, Jal Sathis have been providing better consumer-friendly services because they hail from the local community and are more accountable to their own community.

Women are hit the hardest when there is a water problem at home because, since time immemorial they have been responsible for fetching water for all household needs. Access to clean and safe water at home leads to women's empowerment as women would no longer need to trudge miles and spend hours to fetch water, so we thought it would be most appropriate to have women as partners in drinking water distribution management at the habitation level.

Odisha has a strong legacy of women empowerment, with the Mission Shakti[14] initiative spearheading the women's

[14]Mission Shakti was launched on 8 March 2001 by the CM as a key strategy for achieving women's empowerment through the promotion of SHGs by taking up various socio-economic activities. Nearly 70 lakh women have been organized into 6 lakh groups in the state so far, and a separate Department of Mission Shakti came into existence on 1 June 2021.

SHG movement for the last two decades. CM Naveen Patnaik has always believed that unless women are given a voice and their rightful role in every aspect of life, society and the state can't progress.

Inspired by his vision, in 2001, we federated the block-level and district-level SHGs in Ganjam district and I organized the first women's SHG convention as the collector and invited Naveen Patnaik who had just completed one year as CM.

Today, we have many active and dynamic women SHGs and their federations in urban areas throughout the state. By the time we started upgrading the water monitors to Jal Sathis, the state government had already engaged women SHGs for the collection of electricity dues, paddy procurement and midday meal preparation activities under its Mission Shakti programme.

The H&UDD, through ULBs, has also started partnering with Mission Shakti Groups (MSGs) in several areas, particularly in the sanitation sector (solid waste management [SWM] and faecal sludge management), and achieved impactful outcomes. The MSGs have been successfully managing city-level septage treatment and SWM facilities covering the entire state.

This emboldened us to go for a robust engagement with Jal Sathis with higher roles and responsibilities. We had already tested the effectiveness of the incentive system in other sectors of urban governance, so we followed the same approach with the Jal Sathi model and defined quantifiable performance deliverables and associated incentives.

When we decided to engage Jal Sathis across the state to function as an extended arm of the PHEO, we thought it necessary to set out the guiding principles governing the partnership. A standard operating procedure (SOP) was laid down, which detailed the roles and responsibilities of Jal Sathis and the defining principles of the partnership, to bring clarity to the alliance between the PHEO and Jal Sathis.

The SOP details all the aspects of the Jal Sathi programme: programme objectives, agencies supporting the implementation, implementation strategy, criteria for selection of Jal Sathis, capacity building, roles and responsibilities of Jal Sathis and the supporting agencies, dos and don'ts, incentives, etc.

On 18 December 2019, Chief Minister Naveen Patnaik formally launched the Jal Sathi initiative in a state-level event organized to commence water charges collection by Jal Sathis from the consumer's doorstep using a mobile point of sale (MPoS) system device. By paying the water charges for his residence, Naveen Niwas, and getting a receipt through the MPoS device from a Jal Sathi at the event, the CM became the first consumer to avail of the Jal Sathi service.

The Jal Sathis are now the go-to persons for consumers in municipal wards across the 115 towns in the state for water supply-related matters. Their responsibilities range from testing water quality at the consumer's end to reading the meter, bill generation, delivery, collection of water charges, facilitating complaints redressal and, above all, managing customer relations.

The facilitation role initially played by water monitors has matured into a deeper and more intense role as Jal Sathis, and they have now become the pillars of the urban drinking water management system in Odisha.

The Jal Sathi initiative that started in Bhubaneswar has reached all ULBs across the state, with more than 900 Jal Sathis till August 2022. The selection of Jal Sathis from the women SHGs and their federations is done through a transparent process by the concerned municipalities, and a memorandum of understanding (MoU) is signed between the PHEO and the group/federation.

Each Jal Sathi is assigned a clearly specified geographical area of one or more wards with a few hundred connections having a mix of residential, commercial and institutional consumers to ensure that a decent income is earned by each Jal Sathi.

The Jal Sathis are provided with a uniform to wear while on duty, an ID card, a shoulder bag containing the MPoS device, a water-testing kit and registers. They are also trained in the technical skills required to discharge their functions along with soft skills for customer relationship management before their deployment in the field. Jal Sathis are incentivized handsomely for their services in the water supply management chain.

For every new domestic connection facilitated	₹100
For every new commercial connection facilitated	₹200
For each unauthorized domestic connection that was regularized through the Jal Sathi	₹100
For each unauthorized commercial connection that was regularized through the Jal Sathi	₹200
Field water quality testing (per sample)	₹20
Collection of water charges	5 per cent of the collected amount

Though we are engaging Jal Sathis for various functions on a part-time basis, we expect them to earn a decent income as incentives. To ensure this during the initial period of six months to one year, we paid them a minimum of ₹5,000 each as a monthly incentive in case the actual incentive earned was less than ₹5,000. This was done to enable them to sustain a long-term partnership with the PHEO from the beginning.

Thus, through these part-time activities, each Jal Sathi now earns an average monthly income of ₹10,000–₹15,000 and, in some cases, ₹20,000–₹30,000 per month.

Though the MoU is signed by the respective groups, the incentives are auto-calculated by the system every month based on performance and the incentives are transferred directly to the Jal Sathis' individual bank accounts.

After scaling up to the 115 cities of the state, about 1,000 Jal Sathis will be partners in water supply distribution management, collecting a total annual revenue of ₹300 crore, which would fetch them an annual earning of ₹20 crore.

The Jal Sathi initiative has yielded several significant outcomes in water supply management in urban areas. The notable achievements are:

- 100 per cent HCs
- Improved revenue collection
- Improved customer connect leading to customer trust, confidence and satisfaction
- Direct feedback from consumers on water supply issues
- Customer-friendly water supply services at the doorstep
- Improved awareness regarding desirable behaviours related to water conservation and water supply service
- Elimination of unauthorized connections

The onboarding of Jal Sathis as change agents in pipe water distribution and consumer management has now started giving multiplier effects for the government and Jal Sathis themselves. Due to the economic security derived from the incentivized services that they deliver, the Jal Sathis are now leading a dignified life. Many Jal Sathis have openly stated how this engagement helped them navigate the difficult times during the recent pandemic.

For many Jal Sathis, the incentives are the only source of income for the entire household, which elevates their stature in the neighbourhood and among kith and kin besides earning them dignity and honour within the family. This benefit has been echoed by Jal Sathis in Bhubaneswar and Puri. According to them, their commitment to the programme became stronger as it not only provided economic protection but also gave them social standing.

They are now seen as water professionals in their areas of operation. Functioning as Jal Sathis, these women are witnessing their own personality transformation through economic and social empowerment. The Jal Sathi initiative has transformed the water supply service delivery and taken women's empowerment to new heights. I am very happy that the strategy of onboarding Jal Sathis has become a game changer and that they have now become an indispensable part of the success of Odisha's water supply transformation.

As a token of appreciation for the contribution of the women SHG members in various sectors of urban governance like water, sanitation, etc., and to encourage them further, our department has taken an initiative to provide bicycles free of cost to all community partners including Jal Sathis, to ease their travel requirements.

I can add with pride that inspired by the success of Odisha's Jal Sathi initiative, the Ministry of Housing and Urban Affairs (MoHUA) of the Government of India has included community partnership as an essential component in the implementation of AMRUT 2.0 across all states.

THE ESSENCE

Infrastructure alone cannot bring change. The key to success lies in engaging with communities and empowering them to take charge of their own lives.

—KOFI ANNAN

Community partnership is crucial for successful government project implementation because the community members can provide valuable inputs during the planning and implementation stages, identify potential issues and solutions and help monitor and evaluate a project's progress.

The Jal Sathi initiative in Odisha is an exemplar of the effectiveness of community partnerships. Jal Sathis were selected from the local community and trained to provide customer-friendly services. They became the main pillars of the urban drinking water management system, leading to improved water supply management that achieved 100 per cent HCs and improved revenue collection. The initiative has provided economic and social empowerment to women in Odisha and proved to be a transformative step in water supply service delivery.

Involving community members can foster a sense of ownership and accountability in them, resulting in sustainable success. Additionally, community engagement can help build trust and strengthen relationships between the government and the local community, leading to effective collaboration in other sectors too.

Unleashing the transformative power of community partnership: the game-changing Jal Sathi initiative.

9

LEAVING NO ONE BEHIND

Our water journey, which began in 2015, is marked so far by the following milestones:

- An investment of ₹6,000 crore in infrastructure upgradation between 2017 and 2023 and an increase in annual investments in the urban water supply sector by about five times from ₹200 crore to ₹1,000 crore.
- The water supply distribution network more than doubled from 5,383 km to 11,325 km, with an increase in network coverage from 40 per cent in 2015 to more than 95 per cent in 2022.
- Forty-nine towns that were water deficient in supply, have turned water sufficient with a supply of 70–135 lpcd.
- HCs have increased from 3 lakh to 9.8 lakh, indicating an increase of about 225 per cent.
- Nine water testing laboratories were established across the state to monitor the quality of water supply.

Naveen Patnaik's vision to provide pipe water supply to every house has powered our journey of transforming urban Odisha's water supply scenario from water scarce to water secure.

Strong commitment to the cause coupled with the passionate implementation of the programme, appropriate pro-poor policy measures, innovative process re-engineering, integrity in implementation, community partnership, team building, capacity development, etc. were the reasons for the success so far.

In October 2020, history was created by Odisha when Bhubaneswar became the first city in the country to achieve the milestone of 100 per cent of houses having piped water connections along with 14 other cities of the state. With this milestone, our marathon water journey entered the winning lap and implementation in the other cities gained momentum.

With 105 out of 115 cities achieving 100 per cent HCs by July 2023, we are firmly on track to achieve this target across all the cities of the state by April 2024. When a claim is made that a city has achieved 100 per cent HCs, a question may arise as to how and by whom the claim was verified, validated and certified.

When I asked the EE of a division about how he got the 100 per cent HC coverage status verified for the city for which he was in charge, he didn't have an answer. These questions can be asked by anyone when we claim that a city has achieved 100 per cent HCs; therefore, we decided to put in place a foolproof verification and validation protocol, which could withstand scrutiny.

To this end, a city-level multimember committee with representation from among all key stakeholders, such as the collector's representative, PHEO engineer, city official, councillors, and Jal Sathis, was constituted to undertake joint field inspections. The committee verifies whether the entire city with all the wards, streets, habitations and houses is connected with a piped water supply and that no house is left behind to certify the 100 per cent coverage status.

After a city is certified by the committee for its 100 per cent coverage status, press releases are published in leading local newspapers to invite the public to convey within seven days whether any house is left without a water supply connection.

After providing connections to all the uncovered households reported by the public, the city is declared to have achieved 100 per cent HCs. The PHEO division makes posts in social media and handles tweets about the achievement, which is acknowledged, congratulated and reported by the district collector, EIC, PHEO, H&UDD and, occasionally, the CM too. The journey of a city achieving 100 per cent HCs is considered complete only after the achievement is posted on social media for wider outreach.

We understand that urbanization is a continuous process wherein cities grow every moment. With the constant influx of migrants and the addition of new households, the urban population is dynamic. Though today we claim to be covering 100 per cent of houses with pipe water connections, the situation may change very soon due to the dynamic nature of urbanization.

I understand that claiming 100 per cent coverage is risky due to an ever-growing population and increase in house numbers Consequently, although we achieved 100 per cent on a particular date, someone can prove us wrong in the future by saying that a few houses are yet to be covered. If we had opted for the safer claim of declaring a city as almost 100 per cent or more than 90 per cent covered, it would have given us the safeguard that nobody would be able to prove us wrong, but doing so would have encouraged our team to settle for less than 100 per cent achievement.

Knowing this, we have consciously decided to keep declaring a city to have 100 per cent HCs because it makes the government responsible for looking out for uncovered houses and covering them immediately to make the claim valid at all times. This constant endeavour is intended to achieve and sustain 100 per cent coverage and leave no one behind, thus practising the spirit of the Sustainable Development Goal (SDG) 6[15] of the UN.

This water journey of urban Odisha has brought us several laurels. In 2018-19, the state received the Housing and Urban Development Corporation Limited (HUDCO) award for best practices to improve the living environment. Odisha secured first place for the successful implementation of AMRUT 1.0 across the country for three consecutive years: 2018–19, 2019–20 and 2020–21. In 2020, we also received the award for best civic agency, instituted in memory of V. Ramachandran, by the Janaagraha City Governance Awards.

Several studies across the globe have established that the availability of safe drinking water leads to improved public health. With the state achieving more than 97 per cent HCs across all cities, ensuring access to safe and clean drinking water, we were eager to assess the effects of the renewed and improved water supply system on public health and wanted to empirically verify the effects on the field.

District-wise disease surveillance data were collected from the Integrated Disease Surveillance Programme (IDSP)[16] of the H&FW of the Government of Odisha and analysed. Though the data collated by the IDSP related to entire districts covering urban and rural areas, we were of the view that waterborne diseases generally and predominantly occur in urban areas only.

The data revealed that out of 3,608 jaundice cases and 29 related deaths in 2014 due to the consumption of contaminated water recorded across the state, the numbers

[15]SDG 6 talks about universal and equitable access to safe and affordable drinking water along with access to adequate and equitable sanitation and hygiene for all.

[16]The IDSP is a decentralized, state-based surveillance programme. It aims to detect warning signs of potential outbreaks and helps initiate an effective response in a timely manner.

have consistently declined year after year to 26 jaundice cases with 0 deaths in 2020.

Similarly, diarrhoea cases also declined from 3,832 cases with 30 deaths in 2016 to 146 cases with zero deaths in 2020, a reduction of more than 95 per cent. Though I was happy with the completion of every milestone achieved in the journey, the vast decline in jaundice and diarrhoea cases and the prevention of deaths gave us immense satisfaction and a sense of pride.

When we started witnessing the visible impacts of the journey on public health indicators on the ground, we realized the power of Naveen Patnaik's vision and felt contentment as the very purpose of embarking on this journey was being accomplished.

Having achieved over 90 per cent network coverage and HCs and on our way to reaching 100 per cent soon, the question before us was 'What next?' Most of the pipe water networks established under BASUDHA were nearing completion, and our water journey of providing a pipe water connection to every home in urban areas and leaving no one behind appeared to be nearing its goal by then.

The journey from water scarcity to water security in urban Odisha with 100 per cent HCs gave us rich learnings and enormous satisfaction. At that point in our journey, I had gained a confident team that could take on bigger challenges with more and more Jal Sathis joining the brigade every day; hence, I felt that it was the opportune time to think about moving to higher and more aspirational goals in the urban water supply sector.

10

CLIMBING TO NEW HEIGHTS

During this eventful journey, our team—after gaining experience and rich learnings—evolved into a more competent and confident team that was ready to take up new challenges. While aiming for new horizons for the team to reach, I mulled over the idea of assigning them equally important and still untouched challenges, such as the reduction of water losses, metering connections, improving revenue collections, reducing the exploitation of groundwater sources, migrating to surface water sources and improving water quality, which is necessary to complete water sector reforms and ensure long-term sustainability.

Additionally, the gross disparity in water supply within the cities remained to be addressed. In the meantime, in the year of 2019, sporadic incidences of diarrhoea and jaundice were reported. Though no deaths were caused, those incidents triggered a question: How good is the actual quality of water that we supply?

This kept nagging me. Could we give a guarantee for the quality of water supplied? Could I confidently declare that these reported waterborne diseases were not because of water contamination? I asked my team the same questions that were plaguing me.

'When you claim that good quality water is supplied through the pipe network, why do you continue to drink bottled water instead of tap water?' I asked my EIC during a meeting. All the senior officers attending the meeting looked at each other with a puzzled look on their faces. I then realized that our claim of good quality water supply was more of an assumption than assurance based on verifiable water quality parameters.

Therefore, we decided that besides making an adequate quantity of water available for consumption, we should focus on ensuring good quality water that complied with quality standards. If the supply of good quality water is ensured, we should be able to drink water straight from the tap without further filtering or boiling before consumption.

One prerequisite for preventing contamination in the transmission and distribution network is supplying continuous pressurized water instead of the prevailing intermittent supply. Supplying continuous pressurized water will also result in huge household savings because individual households will no longer require water storage tanks and water filtration systems (RO systems). Inadequate and inferior water quality forces people in the low-income group (LIG) and middle-income group (MIG) to spend a sizeable portion of their monthly household earnings to buy good quality drinking water.

For a majority of the urban poor, including slum dwellers, good quality water is not affordable, and they have to make do with the inadequate and inferior quality of water supplied. After weighing the merits and advantages that would ensue from the 24×7 drinking water supply and the huge public benefits (including improved public health) that it would result in, we decided to set our sights on enhancing the quality of the water supplied.

Though the supply of improved and standardized water quality will benefit the entire urban population, we thought that it would have greater impact on the lives of slum dwellers, urban poor and LIG and MIG people, who constitute the majority of the urban population.

The thought lingering in my mind was that this was a universal problem that people everywhere were facing. It is more relevant for the underdeveloped and developing areas. Ultimately, the poor have to struggle to manage such problems.

Naveen Patnaik always expects only one thing from his team: to do something to transform the lives of the poor. With the enabling government ecosystem and direct support from the chief minister to improve the lives of the people, we marched ahead with our aim to deliver drinking water. If I couldn't solve the water issue in Odisha and that too now, who else would do it and when?

Thus, the 'drink from tap' idea emerged, which soon became the next mission in our water journey: the *Drink from Tap Mission*.

THE ESSENCE

Great things are not done by impulse, but by a series of small things brought together.

—VINCENT VAN GOGH

From a single drop, the vastness of the sea was born: *DRINK FROM TAP MISSION* **in Odisha.**

11

UNLOCKING TRANSFORMATION

Providing a continuous supply of drinking water conforming to IS 10500[17] quality standards to every home, which one can drink straight from the tap, became our focal agenda as we progressed from the universal coverage mission to a higher goal—the Drink from Tap (DFT) Mission.

We were aware that extraordinary efforts would be required to take up this maiden mission due to the very limited, near or nil expertise available in the country. DFT Mission implementation was entrusted to the newly created corporate entity, WATCO.

A core team was formed within WATCO with engineers of the state government who had exhibited extraordinary commitment and passion and had performed exceptionally in the past. The 19-member team for the mission was constituted and notified on 6 August 2019, with Mr P.K. Swain, the then PHEO SE (Design), as the mission director along with other senior officials and a panel of internal and external sector experts as members.

Mr P.K. Swain had earlier exhibited exceptional leadership

[17]IS 10500-2012 specifies the acceptable and permissible quality of drinking water for human consumption in case of the unavailability of an alternative source. This serves as the fundamental standard adopted by the Bureau of Indian Standards for drinking water in India.

and professional competence in completing one of the most complex and unique PPP projects in the drinking water supply sector in the country. He was also instrumental in salvaging the much-delayed execution of the GBWSP. He became my natural choice to drive the mission.

When we were about to move on to higher aspirations in the water sector, Naveen Patnaik returned to power after the general elections held in 2019. With a huge public mandate, he was elected to lead the state for the fifth consecutive term after completing 20 years as Odisha's CM.

At the beginning of his term he decided that governance and development during this tenure must be transformative and not just incremental. He wanted transformative governance to become the unique selling point (USP) of his fifth term, so he introduced the 5T Governance Framework[18]: transparency, technology, teamwork, time and transformation in governance. This decision paved the way for a paradigm shift in public governance in the country. His PS, Mr V.K. Pandian, was designated as the 5T secretary to drive the 5T agenda directly from the CM's office.

Under 5T, all the departments took up many transformative initiatives. The H&UDD took up initiatives under eight broad verticals, of which DFT was one. Mission teams were formed for the eight interventions with senior officers at the special secretary and additional secretary levels heading the missions. The DFT Mission team was also approved to be notified customarily with the special secretary as the chairperson.

However, because DFT is no ordinary mission, I was bogged down by apprehensions about the level of passion and commitment that would be required to accomplish the mission. I was convinced that it would be most appropriate

[18]In 2019, the Government of Odisha introduced and adopted the 5T initiative to provide citizen-centric good governance in all departments.

if I led the mission and gave my absolute best for its success.

Thereafter, I recalled the file in which I had given my approval for the constitution of the team just before the notification was dispatched and changed the order to assume leadership for the mission. The mission team started to intensely deliberate on strategies and the ways to move forward.

During brainstorming sessions, we stumbled upon two more fundamental issues: tackling high water losses and ensuring equity and consistency in the supply of drink-from-tap quality water to every home and from every tap.

Though the demand was at par with water production, from a study undertaken in 2015 we discovered that the unaccounted-for water (UFW) or the non-revenue water (NRW) in Odisha was a whopping 54 per cent of the produced water against the national service level benchmark of 20 per cent. NRW is the water that has been produced and 'lost' before it reaches the customer. Losses can be real losses through leaks (also referred to as physical losses) or apparent losses (through theft or metering inaccuracies).

We firmly decided that the issue of NRW had to be addressed aggressively and urgently to sustainably transition to a 24×7 water supply.

The other issue that needed to be addressed was the installation of personal underground/overhead storage tanks with pumps by individual households, through which water was illegally drawn from the pipe water network during supply hours. This created inequity because households located downstream seldom got adequate water supply.

People tend to store more water than is actually required and waste the excess stored water when fresh supplies are received. The more they store, the more they waste. We realized that by providing a reliable 24×7 water supply. Equity in water distribution and prevention of wastage could be achieved effectively.

We lacked experience in 24×7 water supply systems, so we decided against making a jump start. Instead, we experimented in a small area so that the process could be scaled up to cover cities depending on the outcome. As one of the key objectives was to ensure the supply of safe drinking water to the urban poor, we explored the possibility of selecting a slum area for the pilot implementation.

Under the universal coverage programme, slum households are provided with individual taps in their homes. Due to intermittent and erratic supply, households had to store water in large vessels and buckets, which occupied substantial space in the already cramped dwelling unit. Additionally, the households were subjected to day-long waits and anxiety, which disrupted their daily chores.

This section of the population is completely dependent on the public water supply system, so ensuring good quality drinking water for the slums is all the more important.

We selected a slum as the sub-pilot area to make sure that DFT could be successfully implemented in slum areas, which housed the main intended beneficiaries. It was not intended to jeopardize their health and safety by exposing them to risk during the trials in case of failures.

Had the DFT Mission failed in the slum area where it was experimented, I would have abandoned the mission instead of continuing it in higher-income group habitations where it was easily implementable.

As suggested by the team, on a Sunday morning, my team and I went to Ishaneswar Slum (a small slum on a hillock with 471 households) in Bhubaneswar and climbed the reservoir for a bird's-eye view of the slum. During our interactions with the local residents, we witnessed the precarious and deplorable state of water storage and use at the household level with dismay. We instantly decided to choose Ishaneswar Slum for the implementation of DFT as a sub-pilot.

One of the factors we considered while selecting the sub-pilot area was to have a habitation with limited households and where the water supply could be easily isolated. Thus, any disruptions during the pilot would be limited to a smaller population, which could be managed without causing inconvenience to the larger population.

After the gap assessment study and a detailed asset and consumer survey of the 471 houses, a plan of action was prepared and the idea was finally set into motion. We commenced the work in Ishaneswar Slum on 1 September 2019, with very little idea of the road and challenges that lay ahead.

As the first initiative in the country, we looked for global technical partnerships to bring in expert input and best practices. The United Nations Children's Fund (UNICEF) was identified for technical assistance, and an MoU was signed with them on 25 October 2019. Through the MoU, we received technical support from Mr Richard Franceys and Mr Anand Jalakam, who are renowned for their expertise in 24×7 water supply.

They shared valuable inputs on the learnings, chances and reasons for the failure of 24×7 projects executed across the country. An article jointly penned by them gave details of the shortfalls of and delays in 24×7 projects across India, which gave very useful insights into the factors that contribute to failures and what not to do in 24×7 projects.

The UNICEF experts helped us review the DPR in terms of the design aspects of the 24×7 supply system. DFT was still way beyond the 24×7 system; it required the in-house development of an indigenous system aided by our own hands-on experience.

In November 2019, a selected team of six officials was sent to the Administrative Staff College of India (ASCI), Hyderabad, for a three-day training programme to understand the basics of 24×7 projects. The team returned with new learnings and insights and accelerated the implementation of the sub-pilot work in Ishaneswar Slum.

Meanwhile, the Jal Sathi initiative was formally launched by the CM on 18 December 2019. The Jal Sathi for Ishaneswar Slum was immediately taken on board for the implementation. The move proved to be extremely helpful in the successful implementation of DFT in Ishaneswar Slum.

The DFT Mission in Ishaneswar Slum was completed, commissioned and operationalized on 31 December 2019, thereby meeting the deadline. Immediately after commissioning, the team faced many challenges, such as endless leakages and unexpected surges in water demand. Unperturbed by the challenges, the team addressed them one by one and tied all the loose ends. In a month, the system started to stabilize.

To further sharpen our strategy and ensure that we were on the right track, we reached out to the globally renowned water expert Dr Asit K. Biswas, former professor at the National University of Singapore and a Stockholm Water Prize winner. We sought strategic clarity on the operation and sustainability of the 24×7 water supply system.

As suggested by Dr Biswas, a six-member team from the DFT Mission was deputed to Singapore in January 2020 for week-long training at the Singapore Water Academy (SWA) of the Public Utilities Board (PUB), Singapore's National Water Agency under the Ministry of Sustainability and the Environment (MSE).

Immediately after their return, a one-day workshop was organized on a Sunday to harness the newly acquired knowledge and insights of the trained officials. The team was asked to present their learnings and takeaways from the SWA training and draw up a DFT implementation roadmap for the state. All senior-level engineers and divisional EEs from all the PHEO divisions of the state were invited to attend.

After the team presented their learnings, the roadmap with short- (next six months), medium- (two years) and long-term (past two years) strategies for DFT implementation in Odisha was drawn up with 28 action items and the responsibility matrix and timelines.

The roadmap became the guiding document for us in the following days. This seamless process of making the team present their learnings and creating a roadmap for implementation immediately upon their return from training abroad ensured that there was no gap left between learning and doing. The workshop was a watershed moment of the DFT Mission; therefore, for me, that Sunday turned out to be an unforgettable day.

THE ESSENCE

The secret of getting ahead is getting started. The secret of getting started is breaking your complex overwhelming tasks into small manageable tasks, and starting on the first one.

—MARK TWAIN

With a strong desire in our minds, we embarked on the DFT journey without any readymade blueprint. This forced us to learn the hard way through trial and error. Despite receiving discouraging feedback, we were determined to create a roadmap for future generations.

Any significant achievement requires taking the first step: the journey towards success may be long and challenging, but perseverance and dedication can lead to great accomplishments.

Odisha's journey towards transforming its urban water supply system into DFT started with a single step of piloting the mission in a small slum of about 500 households. It was then upscaled to 12 pilot zones, one of them being the largest slum of Odisha, Salia Sahi, followed by covering all of Puri and then 22 more cities with a target of 23 DFT cities in Odisha in the next 15 months.

By starting small and gradually scaling up, it becomes easier to identify and address challenges, thus ensuring the success of the project in larger areas. In the sprawling slum of Salia Sahi, the implementation of the DFT Mission faced several obstacles from limited water infrastructure to opposition from antisocial elements and water vendors. However, the DFT team's citizen-connect approach overcame these challenges, and with persistent efforts, they provided piped water to all households in the area. The success of the project hinged on social engagement rather than physical infrastructure, with two taps per household becoming a significant empowerment initiative for women and improving health and hygiene.

Using the systematic step-by-step strategic approach, we piloted the mission in a smaller area, faced challenges, addressed them and stabilized the system in a month. This led to the creation of a blueprint for DFT implementation in Odisha, which was a watershed moment in the journey.

Lab to land: experimenting with DFT with a step-by-step approach.

12

LAB TO LAND

Ishaneswar Slum holds a special place in the journey of the 24×7 DFT Mission in Odisha. It served as the mission's foundation and emboldened us to upscale it. DFT implementation in Ishaneswar Slum brought us instant success and enriched us with enormous learnings.

While it gave us strong insights into our assumptions, some of which proved to be right and many proved to be wrong, it also brought a lot of unforeseen challenges that were resolved during implementation. The experience gained in Ishaneswar Slum helped us develop a broad outline for planning, designing and upscaling DFT projects.

While selecting the pilot area and during the execution of the pilot, the team had set its eyes not only on success but also on all probable issues and bottlenecks that might crop up, to devise a dynamic strategy that could take care of the challenges likely to be encountered in the scaling-up phases of the DFT.

Had this not been done, the team would have found it extremely difficult to scale up the mission in Puri because the magnitude of unforeseen challenges would have been very high for the team to tackle, besides affecting the schedule for project completion.

We kept reminding ourselves that failure was not an option. Thus, the numerous challenges encountered during sub-pilot implementation seldom dampened the spirit of the team, and we were eager to experiment with solutions for all possible challenges in the sub-pilot itself.

After the sub-pilot construction was completed, we were startled to know that the slum with 471 households consumed more than double the standard quantity of water normally supplied to 471 households.

Because we knew that water demand and consumption at the slum-household level could not be so high, we decided to engage a team from the Indian Institute of Technology (IIT) Madras to probe this abnormality in the consumption pattern. This team deployed closed-circuit cameras, ran leakage checks in all the pipes and started plugging the leakages.

We soon found out that the major reason for the abnormal consumption of water was large-scale leakage in the pipe network at the ferrule connection points of all households. The house lines were connected to the street pipe network by ferrules, and the ferrule joints often became loose because of heavy vehicular movement in the streets. Because of loosened ferrules, every point connecting the street line with the 471 household connections was leaking.

Due to gravity, the water leakage dribbled downward and got absorbed into the earth, leaving no visible evidence on the road surface.

Although the leakages at the individual HC ferrule points were seemingly insignificant in volume, the gross leakage at all ferrule points and at all times of supply resulted in huge water losses.

Furthermore, all these leaking joints acted as potential points of contamination, with impurities surrounding the leakage points getting sucked in due to reverse pressure whenever the supply was stopped. It dawned on us that resolving this ferrule joint issue would help us significantly reduce water losses and contamination.

We resolved the leakage problem by replacing the

faulty ferrule connections with new ones along with saddle and compression fittings. This intervention paid off. After monitoring the piped water supply for a month, we noted that the quantity of water lost due to leakages had dramatically decreased.

We also realized that the ferrule point leakage issues were attributable to the prevailing connection policy. Under the Odisha Water Rules, 1980, the responsibility of having an HC was placed on the consumer. Consumers were required to get the HC executed at their own cost by engaging a plumbing contractor and procuring the required pipe and other fittings, so most of them used cheaper material (mostly non-standardized and low quality) and engaged untrained and unlicensed plumbers who would do the work at a low cost.

Though the policy of making the consumer responsible for the HC was to minimize the government's spending on water supply work, we realized that by doing so, we had seriously jeopardized the safety of the water supply network system and compromised water quality.

The government also ended up spending more than what it would have otherwise saved by providing HCs on account of recurring water losses and the risk to public health due to the poor quality of water supplied.

Having understood the dynamics and economics of HCs, we decided to address the root cause of the issue and amend the policy by declaring HCs as public work and liberating the consumers of the responsibility.

With this paradigm shift in policy, the HC process was streamlined and an SOP was notified that stipulated the connection technique using a saddle with compression fitting, standards of materials to be used, workmanship, etc., to ensure leakproof connections to households.

Estimates determined by the team reflected that the

average execution cost per household would be approximately ₹3,600. We decided to peg the charges at ₹3,600 for each HC and to collect the charges from the consumer either as a one-time lump sum payment or in smaller instalments.

However, considering the struggles of the urban poor in making ends meet with their paltry household income, I wanted the state to exempt the urban poor from paying the connection charges and to provide them with no-cost connections. When this proposal was submitted to the CM for approval, he approved it and sent it back the same day without asking any questions or seeking a discussion, even though it burdened the exchequer.

This is how the constant encouragement and push from CM Naveen Patnaik for pro-people reforms facilitated the journey in bits and pieces. Thanks to his benevolence, these transformative policy interventions resulted in hassle-free HCs for consumers, the speedy execution of HCs, prevention of leakages and contamination-free connections. I am happy to add that this game-changing policy decision was later recognized as a best practice by MoHUA of the Government of India.

For the implementation of DFT components, such as metering, leakage reduction and revenue collection, community participation and involvement are vital. Our indigenous concept of deploying Jal Sathis for community mobilization reaped rich dividends and ensured the successful implementation of the mission.

Generally, accountability for consumption induces responsibility in the minds of consumers. Unless a control mechanism that establishes accountability for usage is implemented, it is not possible to keep the leakages and wastages in a public distribution system under check. Therefore, we decided to introduce metering as an essential component under the DFT Mission.

Metering the connection is also seen as a tool to inculcate responsible behaviour in the consumer. We were able to establish that metering instils accountability and responsibility in both the user and the supplier.

Before the pilot in Ishaneswar Slum, water meters were not used in the state. Although a minuscule pilot in the past had attempted installing water meters, it was presumably not taken forward. I won't be wrong in claiming that Ishaneswar Slum was the first area in the entire state to have water meters.

Metering consumption is essential for ensuring the sustainability of DFT projects, so we made it an integral part of the DFT Mission. However, selecting the appropriate meter that would suit our requirements took us some time because metering was being adopted for the first time in Odisha.

We were swamped by innumerable agencies showcasing different meters that used the latest technologies and provided a wide range of benefits. There were analog, digital and radio frequency identification (RFID) meters or smart meters, and the range of fascinating benefits offered by high-end meters was very tempting.

We had detailed sessions discussing the basics of what kind of meters would meet the requirements and the economics of procuring them. Our team critically analysed the pros and cons of several types of meters, and we consciously decided on the most basic mechanical meter.

Here, I would like to recall an incident that is vividly etched in my memory. Water metering had always been a priority for us that awaited implementation as part of our water sector reforms. Based on our inputs on the desirability of water metering, the 4th State Finance Commission (FC) made a specific recommendation for water metering and provided ₹375 crore exclusively for this purpose.

We were eager to start implementing water metering in the state, and when the first State Annual Action Plan (SAAP) under AMRUT 1.0 was considered in 2016, we decided to include water metering as one of the projects. A proposal was moved to approve the procurement of smart water meters. Though smart meters were very expensive (about ₹18,000 in 2016) compared with mechanical meters (about ₹2,000), the technical committee chose smart meters considering their fascinating features.

In the afternoon of the following Sunday, I received a phone call from a Member of Parliament (MP) enquiring about the water metering proposal and suggesting a reconsideration of the decision to procure smart water meters due to their high costs. The fact that a politician could come up with a suggestion on a decision taken by a state-level technical committee headed by me was not very palatable to me.

Nevertheless, I decided to give it a thought and withheld the decision pending its re-examination. The life expectancy of the smart meters proposed to be purchased was about five years, and the average water charges collected per month ranged from ₹150 to ₹250. Upon reviewing our decision to go for smart meters, we found that it would take about seven years to fully recover the cost of the meter, that is, about ₹18,000, from the water charges collected, which was two whole years after the expiry of their life expectancy.

Though smart meters have smart features and use the latest technologies, they will be suitable only for sectors with a higher tariff base, such as energy, and may not be economically viable for systems like ours with low-tariff structures.

The technical committee decided to push water metering projects for consideration at a later time with suitable modifications and decided to focus on implementing universal coverage projects instead. The key takeaways from this experience were:

1. To err is human, whether it's a technocrat or bureaucrat. Though our intentions for water metering

with smart meters were bona fide, we missed weighing the financial prudence of selecting high-end meters for low-tariff applications. Any decision must adopt an all-encompassing approach, and a decision-maker should not get carried away by fascination.

2. Every opinion matters. Insights can come from anyone, and we need to be open to suggestions. Though I had felt uncomfortable receiving a suggestion from the MP who questioned the decision taken by a technically competent team, the idea of giving it a thought and revisiting our decision helped us avoid a costly blunder.
3. Free and open communication must be encouraged within the system. I presume that the MP was briefed by one of our own officials who had difficulties in sharing his views with us out of fear of reprimands from the seniors. Fortunately, the MP magnanimously relayed the valuable suggestion, which prevented us from going ahead with the unviable proposal.

Because the Cabinet had already approved the implementation of the 4th FC recommendations, including water metering, in toto, we moved a proposal to defer water metering till the time the universal coverage of tap water connection was achieved.

Cabinet approval was obtained for our proposal to utilize the funds earmarked by the 4th FC for water metering in water supply projects. We were very clear about our objectives and priorities and did not digress from the set path. No stone was left unturned, and all the stumbling blocks were overridden.

The decision to use basic mechanical water meters instead of smart meters in the DFT Mission later proved to be a smart move. Our objective was to introduce the metering of water consumption to give a clear message to the consumer that their consumption would be measured and that they would have to pay for the consumption, including leakages and wastages, at their end.

With the introduction of metering, positive behavioural changes were witnessed in consumers and their consumption patterns became more disciplined. Given their reasonable accuracy, mechanical meters have served their purpose of regulating consumption.

Ishaneswar Slum served as a living laboratory for us in DFT implementation and helped us develop a basic yet full-fledged implementation framework, which we piloted in a few zones in Bhubaneswar and Puri for full-scale fine-tuning prior to upscaling. We ensured a systematic approach with steady learning and doing curves moving from the laboratory to the ground. Eight pilot zones in Bhubaneswar and four in Puri were selected.

As part of the implementing strategy, we selected areas that fairly represented the diverse urban fabric by means of a mix of habitations, including slums, low-income houses, middle-income houses, high-income houses, commercial and institutional areas, and government buildings and quarters.

The pilot zones in Bhubaneswar included Salia Sahi, which has a population of 57,500 people. By selecting Salia Sahi, we undertook the challenge of implementing DFT in the most difficult area of the state.

Salia Sahi is the largest slum in Odisha and is located in the north-central part of Bhubaneswar. Spread over around 252 acres of land under Ward Nos 16, 20, 21 and 26 of the Bhubaneswar Municipal Corporation (BMC), it is a conglomeration of 36 slum clusters surrounded by landmark structures and areas, such as Fortune Towers, the Xavier Institute of Management, the Gopabandhu Academy of Administration, Ekamra Kanan, Loyola School, Hotel Mayfair and Hotel Trident.

Before the DFT Mission's intervention, groundwater had been the main source of water to Salia Sahi, which was accessed through open wells, hand pump tube wells and public stand posts (PSPs). Whenever required, water tankers supplied water to scarcity pockets. A few *anytime water machines* (ATMs) also allowed the dwellers to buy water in times of need. However, these ATMs elicited little interest from the residents due to the price tag and soon became inoperative.

While most of the infrastructure was provided by the state government, WATCO or the municipal corporation, some was created by the residents. Many of the hand pump tube wells were functional but some were not. Many of these had no proper drainage facilities, thus paving the way for water stagnation and associated health issues.

In some places, PVC storage tanks had been installed on elevated pedestals and were periodically fed by the production wells or tankers. Some individual houses in these neighbourhoods had made *jugaad* arrangements by connecting the storage tanks to their individual houses through PVC pipes.

Water was supplied for less than two hours to the houses in the area. For all practical purposes, drinking water infrastructure work commenced in this area for the first time with the implementation of AMRUT 1.0. The execution of projects under AMRUT 1.0 was then moving at a snail's pace and was saddled with various challenges.

Initiating DFT work in Salia Sahi required fast-tracking the execution of ongoing AMRUT 1.0 work and a wide array of interventions, such as infrastructure gap assessment, asset and consumer mapping (geographic information system [GIS] based) and designing the 24×7 system.

The execution of a typical water supply project had always been a limited affair among the contractor, engineer and a few locally influential people who called the shots. The community connect component never found a place in the traditional work culture of the PHEO. The local community and the public at large were seldom involved in the process.

Contrary to the prevailing system, the DFT Mission focused more on citizen connect, which had gained significance in Salia Sahi. During implementation, some unscrupulous elements, who had been indulging in nefarious activities in the locality, felt threatened by the presence of officials in their area. Some others in the neighbourhood who were in the business of selling water to the households also opposed the implementation of the DFT Mission.

Because the mission to cover all households with piped water posed a huge threat to the earnings of the water vendors, they tried to sabotage the execution of the work by planting vicious stories and spreading rumours among the inhabitants about the probable levy of higher user charges for the water received, houses getting damaged by the project work, damage to the roads and pavements caused by the pipeline and plumbing works left unrepaired. They tried to create a frenzied mob against DFT implementation.

During the early days of the pilot project, our team members could not even enter Salia Sahi, as the space was predominantly controlled by antisocial elements. However, we made relentless efforts to get a foothold inside the slum with slow, measured and persistent steps.

The team members concentrated on increasing their interaction whenever they entered Salia Sahi. Though it took considerable effort, we were able to establish a positive rapport with the residents and initiate the work on DFT within a short span of time.

The lack of information and low awareness levels of Salia Sahi residents created issues in garnering local public support at every stage of implementation and necessitated

IEC activities to bring the residents on board much before the actual commencement of the work under the mission.

The team devised IEC activities that focused on the objectives and plan of implementation of the work in the area and highlighted the benefits that would accrue to the residents against those received from the existing facilities.

A communication strategy[19] was prepared with technical support from the International Resource Centre (IRC) and UNICEF. Several group discussions, often in the presence of the local community leaders, were conducted. Interpersonal contacts and communications were established with household members, especially women, mainly with the help of Jal Sathis of the area.

Posters, banners and billboards depicting the message of the mission were installed at prominent places like road junctions and community centres within the settlement. Leaflets and brochures on the proposed system were distributed among the literate section of the settlement. Mobile IEC vans decorated with key messages on the benefits of the mission on both sides and the rear coupled with audio messages criss-crossed the streets and lanes to make an impact on the residents and encourage them to avail themselves of the facilities being extended by the mission.

Salia Sahi is a very dynamic slum set-up with undefined boundaries, lanes and by-lanes. Therefore, the team had to undertake a detailed survey of the boundaries, roads, assets and houses. Due to frequent changes of residence by slum dwellers, in the case of migrant labourers, ascertaining the identity of occupants was challenging.

To begin with, a 100 per cent detailed household survey was conducted to ensure universal coverage. Conducting a

---

[19]The communication strategy was released by the Honourable CM of Odisha on 13 October 2020 while launching the DFT Mission for the state.

household survey in such a thickly populated area proved to be an arduous task. Identifying and demarcating the houses separately from each other is often a matter of confusion, especially in very narrow lanes and by-lanes where clear-cut boundary separations between houses are non-existent. On-the-spot meticulous observation along with detailed local enquiry helped us survey the area to a large extent.

The survey revealed 12,055 houses located in the three zones of the area. The team had difficulties in covering locked houses and households in which occupants were not present during the survey as well as during the execution of the work, which hampered programme implementation. The area, being a conglomeration of urban poor settlements, has a considerable floating population of daily labourers.

Most members go out to work for the whole day and leave their houses locked and inaccessible. One section of the floating population is the farm workers who come from villages during non-farming seasons in search of work and stay in the settlements for a few months. These houses normally remain locked or vacant, making the planning and execution of the programme difficult.

However, through repeated visits, especially during the hours that were convenient to the households that remained closed during the day, the project team members resolved the issue to a large extent. Undaunted by the innumerable challenges, the team adopted several strategic measures to execute the work.

Developing the water supply infrastructure for Salia Sahi included laying distribution mains in uncovered areas, covering narrow lanes and by-lanes, fixing leaky ferrule connections, setting up systems in kuccha dwellings, laying pipelines under concrete roads and pavements, constructing ESRs, ensuring 100 per cent connection to all households, metering the connections and installing chlorine dozers and analysers using programmable logic controllers for real-time monitoring.

Concurrent with the laying of distribution pipelines with their accessories, all infrastructure assets in the area—existing and new—were mapped on a GIS platform for use during execution as well as O&M.

Although the contours of the DFT work at Salia Sahi were similar to those of Ishaneswar Slum, the process, strategy and methodology required for the areas differed. The experiences of Ishaneswar Slum and Salia Sahi taught us about the two broad aspects of the DFT Mission. One is the engineering works, which existed below the ground, while the other is social engagements which existed above the ground.

The former normally remains constant, whereas the latter is extremely dynamic, and more often than not, the component above the ground determines the success and sustenance of the project implemented. For this very reason, I truly consider DFT to be more of a social intervention than physical infrastructure (physical, technical and pipeline infrastructure) because we get to deal with the consumers and habitats.

Thus, DFT is actually a confluence of various streams, parallels and themes. When a pilot is implemented in a government area, the community interface will be minimal, whereas, in a typical slum area, the human interface will be at its highest because of the density of the population and households.

For example, I can boast of one of the earliest reforms made way back in 2016–17 under AMRUT 1.0. Herein, we took an extremely bold decision to provide two taps per household, one in the kitchen and the other in the toilet, for urban poor households as part of the project with government investment without waiting for the finance department's approval.

To me, this was real empowerment, especially for the

women of the households because the kitchen tap liberated them from the burden of fetching water from outside. Further, the provision of a tap in the toilet helped curtail open defecation habits and ensured enhanced health and hygiene. This revolutionary decision of providing two taps to the urban poor consumers' households free of cost was extended to all the projects under AMRUT 1.0.

Had I resorted to the normal channel of seeking approval for providing two free taps to the consumers, the proposal would never have seen the light of day with the concerned authorities preferring to escalate the issue to a higher level for a decision. I thought that going ahead with the decision without seeking approval was the right way of doing it, and it turned out to be a great empowerment initiative.

I was conscious of the fact that I could make such a bold decision at my level only because I was confident that the CM would not hesitate to support bold and fast decision-making if it was bona fide and taken in the public interest.

In Salia Sahi, though we had sanctioned two taps per household, most houses had neither a kitchen nor a toilet. Slum dwellers have just one small room, and if a tap was provided inside the house, it would only add to the inhabitants' miseries with the floors and walls of the tenements getting wet. Hence, in most of the houses, the taps were fixed outside on the front wall.

The pilot implementation of DFT almost completely coincided with the COVID-19 pandemic. The lockdown was declared in March 2020, by which time the sub-pilot in Ishaneswar Slum was complete. The pilot actually started during the pandemic. While the entire country came to a standstill with the national lockdown[20] that imposed lots of

---

[20]Prime Minister Narendra Modi called for a complete lockdown of the entire nation beginning on 24 March 2020 in an effort to contain the COVID-19 pandemic in the country.

restrictions on public movement, the DFT construction work was progressing at a fast pace.

Ironically, the lockdown proved to be a blessing in disguise for the pilot project. The water supply workforce was categorized as emergency workers, and we got special permits to work during the lockdown. Because there was no traffic on the roads and the workforce was sourced locally, we could uninterruptedly dig roads, lay pipelines and quickly restore the roads.

However, during the lockdown, our workforce faced difficulties in gaining access to many places because of the threat of COVID-19. The residents were extremely reluctant to allow outsiders fearing that they might turn out to be carriers of the dreadful virus.

Many localities put up barricades in the streets and lanes to restrict entry for outsiders. Refusing to give up, our team's perseverance helped us enter the project areas under the guise of a water supply maintenance team and carry out the execution process without much ado. We then capitalized on the lockdown restrictions and completed the entire work within six to nine months, which would have otherwise taken a minimum of twelve months to complete.

With the completion of all 12 zones in the pilot phase, we decided to launch the DFT Mission. A state-level event was organized on 13 October 2020, with the CM as the chief guest and dignitaries such as the MoHUA secretary of the Government of India, chief of UNICEF India and chief of WaterAid UK gracing the occasion. The DFT was first formally commissioned in the country in Salia Sahi, the largest slum in Odisha, and it played an extremely important role in our DFT journey.

At the event, the DFT facilities in the 12 pilot zones were dedicated to the two lakh residents populating that area. The event also saw the commencement of the upscaling of

DFT to the remaining areas of Bhubaneswar and Puri to cover entire cities.

This dedication of DFT in the 12 pilot zones in Bhubaneswar and Puri by the CM was a defining moment in our DFT journey, which gave us a lot of pride and happiness. From the time we began our DFT journey till we completed all 12 pilot zones, we were a little apprehensive about the outcome and were not confident about making the project and its progress public.

I met Mr Pandian, PS to the CM, on a Sunday morning at his residence and told him about our DFT experiment. I told him frankly that it was a very challenging task and that the outcome couldn't be guaranteed. He listened to me patiently and ultimately said, 'Don't think about failure. Go ahead. It is worth attempting. We are going to succeed.' His words were a big morale booster for my team and me.

With the completion of all the pilot zones and the smooth operation of the system for a few months thereafter, we gained the confidence to go public with our DFT Mission. Traditionally, at a launch event, the foundation stone is laid and work is commenced. However, the DFT launch event defied this norm; instead, it saw the dedication of all the 12 pilot zones and the commencement of the upscaling of DFT in the other areas of Bhubaneswar and Puri. Our Puri journey began at that point in time.

NOTE: 24×7 is an engineering intervention, whereas DFT is social transformation. Social transformations are more sustainable than engineering interventions. It ensures equitable access to water apart from other unique features.

THE ESSENCE

Coming together is the beginning. Keeping together is progress. Working together is success.

—HENRY FORD

Water supply projects are typically viewed as engineering feats, but my experience has shown me that it's a complex socio-engineering endeavour at least in the context of the developing world. Although engineering plays a critical role in water supply projects, it is not the sole factor that determines success. Various other elements, such as strong political will, proactive administrative machinery and community participation, must come together for any programme to succeed.

Public water supply projects are designed to serve a diverse range of people, which presents numerous challenges in their implementation. From the GBWSP to the BBWSP, I gained first-hand experience in navigating the complexities of water supply projects and witnessed the challenges of acquiring private land for laying pipelines, the importance of a well-structured policy and the benefits of PPPs.

One of my key takeaways is the need for a holistic approach to public water supply projects. It's not only about laying pipelines and building WTPs but also about understanding the social dynamics of the communities that will benefit from these projects. It's about building trust and forging partnerships among government agencies, private investors and local communities. Only then can we achieve a sustainable and equitable water supply for all.

Getting it right by unlocking the power of social engineering.

13

DRINK FROM TAP

The experiences gained during DFT implementation in the 12 pilot zones of Bhubaneswar and Puri gave us the confidence to go beyond the pilot areas to cover entire cities and resolve to take the DFT Mission to the next level, i.e., upscaling at a pan-city level. We were contemplating initiating citywide implementation in Bhubaneswar and Puri, where we had successfully completed pilot implementation.

Though Bhubaneswar would require more funds, a larger trained team and a longer time than Puri, I was biased towards Bhubaneswar becoming the first city to have citywide DFT. The reason is that Bhubaneswar is the capital city with a population of more than a million, so successful implementation could be more impactful and bring mileage to the project. Moreover, I would be able to closely drive the implementation on a daily basis while being stationed in Bhubaneswar.

I discussed my plan of citywide DFT coverage and upscaling in Bhubaneswar with Mr Pandian and sought his consent. He thought for a while and firmly said, 'No, sir. It should be implemented in Puri first.'

He proposed Puri because, at that time, the state government had taken up the Puri temple area transformation project, which was aimed at resurrecting Puri as a global heritage and spiritual centre. He thought that making

Puri a DFT city would add great value to the overall Puri transformation initiative.

Puri has been the spiritual and cultural capital of Odisha, and the people of the state have a unique emotional connect with the city. Puri transcends all regional perceptions and biases; hence, it has great unifying value within the state. Additionally, Puri is the abode of Lord Jagannath and has tremendous emotional influence on all Odias worldwide.

Because of the above-mentioned reasons, Mr Pandian was in favour of starting the first citywide DFT implementation in Puri rather than in Bhubaneswar. In hindsight I must admit that he was absolutely right in doing so.

Had we taken up the implementation in Bhubaneswar or any other city in Odisha, we would not have gotten the recognition and mileage that we received from Puri. Puri is a heritage city and is one of the Char Dham[21] (four holy places) visited by over 20 million tourists annually. Its selection as the first DFT city added greater national and international visibility to the DFT Mission. It would have been difficult to get such visibility for the success of the DFT Mission in any other city. Thus, Puri acted as a brand ambassador for Odisha's DFT Mission.

Mr Pandian's decision to select Puri became a masterstroke in DFT Mission implementation as it brought global visibility and enormous goodwill. Implementation in Puri was started on a war footing despite COVID-19 pandemic-related restrictions. At the same time, implementation in Bhubaneswar continued.

As part of the pilot implementation, 4 out of 19 zones were already covered in Puri. The infrastructure assessment

---

[21]The four Dhams are Badrinath, Dwarka, Puri and Rameswaram. Defined by Adi Shankara, these are four Hindu pilgrimage sites, and it is believed that every Hindu should visit the four Dhams during their lifetime.

and gap analysis for DFT implementation in the remaining 15 zones were carried out, and the DPR was prepared in-house. The water supply infrastructure, including the WTP in Puri, had been built over a long period of time, with substantial investments under the Jawaharlal Nehru National Urban Renewal Mission (JNNURM), AMRUT 1.0 and state funds.

Furthermore, substantial infrastructure improvements were made during Nabakalebara[22] in 2015. Consequently, most of the water supply infrastructure was fairly new or upgraded and in good condition.

Transitioning from the existing intermittent water supply system to the continuous supply of DFT-quality water was the challenge. Tenders were invited to select the implementing agency in December 2020 with a target to complete the work by July 2021.

We were disappointed when the bids were opened and the bidder was finalized. The contractor who quoted the lowest price and met the tender conditions was not known to be very competent in completing work on time.

The chief executive officer (CEO) of WATCO, Mr P.K. Swain, was visibly upset with the selected bidder. He was concerned about delays by the contractor in completing the mission in Puri by July 2021 because he was determined to see the completion of the project before his superannuation, which was due in August 2021. Cancelling the tender and retendering was not an option because the selected bidder was qualified and any retendering attempt would further delay the implementation process.

Hence, Mr Swain decided with reluctance to work with the selected bidder, but he firmly resolved to complete the

---

[22]Nabakalebara is the ritualistic recreation of the wooden icons of four Hindu deities (Jagannath, Balabhadra, Subhadra and Sudarshana) at the Jagannath Temple, Puri. It symbolizes the demise and rebirth of Jagannath and his siblings.

project on time at any cost despite this contractor. From day one, he started closely monitoring the work output and shortfalls and prepared the team with alternative arrangements to make up for any shortfall by the contractor. In a way, his approaching superannuation helped complete the project in record time, as he was determined to inaugurate the project before his retirement.

Ensuring the stakeholders' confidence and participation was critical for the uninterrupted, speedy and time-bound execution of the work. Given the COVID-19 restrictions, limiting the movement of people, materials and machinery while ensuring speedy execution turned out to be very challenging. The district administration assumed greater significance in facilitating uninterrupted field implementation despite the pandemic restrictions and in soliciting community support.

Executing the DFT Mission, which involved the large-scale digging of roads, diversion of traffic, inconvenience to inhabitants and presence of a large number of commercial establishments, including hotels and shopping complexes, warranted the involvement of various stakeholders, such as the district collector, superintendent of police, temple administration, priest association, city administration, hotel and trader association, and public representatives.

Dealing with multiple stakeholders made the implementation process extremely complex and sometimes very frustrating and challenging. In this context, the collector and district magistrate, who is also the administrator of the Puri Municipality and a member of the managing committee of the Jagannath Temple, played a crucial role in ensuring coordination among various agencies and interest groups. The overall leadership role given to him proved to be the right move for ensuring effective coordination.

The Jal Sathi initiative, which was started in Bhubaneswar

in December 2019, was extended to Puri during the pilot phase itself. Encouraged by the advantages of community connect through Jal Sathis, we decided to onboard one Jal Sathi for each ward during the planning stage itself, which proved to be crucial to our success.

Our fears came true, and the contractor started failing to achieve interim milestones. Our team was prepared with alternative arrangements, so local vendors or agencies were plugged in as and when necessary to complete the shortfalls.

The contractor was made to pay for such work by invoking the contractual provision of remedial measures to complete the shortfall in work at the contractor's risk and cost. Efficient contract management practices were adopted without any loss of time and with a view to completing the work within the stipulated timelines.

WATCO's Puri team was headed by Mr Sarat Mishra, the then general manager of WATCO, Puri Division, who is currently serving as the chief operating officer (COO) of WATCO. He was supported by two assistant engineers and six junior engineers, with a few posts remaining vacant. Onboarding the 32 Jal Sathis made a big difference by making the team stronger due to community connect and became a game changer in the success of the DFT Mission in Puri.

The team worked tirelessly round the clock with full commitment and enthusiasm and ensured the completion of the DFT Mission in Puri one month before the deadline. The laying of several kilometres of pipeline networks, construction of new ESRs, achieving 100 per cent HCs (32,017 connections, including laying new and replacing existing HCs), 100 per cent metering of HCs and other ancillary works were taken up to transform the existing system into a complete 24×7 DFT system.

Puri was fully dependent on groundwater for drinking water for decades. Puri's groundwater source was also limited to two sweet water zones[23]: one in Baliapanda and the other at Chakratirtha Road. Being a coastal city, Puri received rapid ingress of saline water, so groundwater sources were depleting fast.

On the direction of the High Court of Odisha based on a writ petition, the Government of Odisha imposed a blanket ban on all kinds of construction and development work in the two sweet water zones in 2000.

The drinking water supply was under threat due to the depleting groundwater in Puri. A project was initiated under the JNNURM to shift the source from a production well to a surface source at the Bhargavi river, 15 km from Puri.

Due to land acquisition-related issues for the construction of an earthen storage reservoir, the project was delayed, then continued under AMRUT 1.0 and completed in 2017. Upon completing the migration of the water source, another project was taken up under AMRUT in the sweet water zone to recharge the groundwater with treated water from the river source.

This was an attempt to replace the water drawn from the earth since time immemorial. Soon after the supply of water from the river source was stabilized, we decided to close all the production wells in the sweet water zones and fully migrate to the surface source.

Today, Puri is completely river-fed without exploiting groundwater for the public water supply system. With 100 per cent piped water HCs achieved in Puri, we can confidently say that Puri has fully migrated to a surface source in the true sense, which is an important milestone in achieving water security.

---

[23]The Odisha Coastal Zone Management Authority declared Baliapanda (207 acres) and Talabania (448 acres) as specially protected areas in 2017 and placed several restrictions against drawing water by sinking borewells and tube wells.

It was exciting for the entire WATCO team to complete the work and commence the trial run of the DFT system covering all areas of Puri. Each day of the trial run was a challenge with surprises, which the team tried to quickly understand and address. Multiple teams were working on multiple work fronts throughout the city to resolve those issues. Door-to-door visits, gathering inputs from consumers, identifying the problems and fixing them and cross-learning among the team members were some of our routine activities during the trial run.

Initially, the demand for water was high for a month or so, and after the leakages were rectified and intensive awareness was created through Jal Sathis on reducing water wastage, the demand gradually came down as anticipated.

Factors such as the 24×7 availability of water, installation of water meters, payment of water charges as per consumption and one-on-one awareness creation by Jal Sathis at consumers' doorsteps collectively contributed to reducing consumption and wastage.

Once all the households were provided with HCs, it was decided to remove the PSPs (street taps) to prevent water wastage. The staff was initially reluctant because stand posts had existed for a long time and provided free water for everyone; they also symbolized the existence of a public water supply system in the city.

As cities do not reach 100 per cent HCs, the abolition of stand posts was seldom attempted and hence appeared strange. Nevertheless, I was fully convinced that it was a logical step to abolish stand posts once 100 per cent HCs and 24×7 supply was achieved at the city level. It was a tough stance taken by me.

Once all the houses got individual connections, there was no need for a PSP in residential areas. Street taps were now required and justified only in floating public areas or high-footfall zones. This abolition of stand posts was implemented in residential areas only, whereas newly upgraded water faucets were installed in public areas, such as commercial streets, tourist places and streets surrounding the Jagannath Temple.

There were about 1,110 PSPs across Puri. Initially, we faced huge public resistance against their removal because the people felt that stand posts function as safety nets and a fallback system if household supply is disrupted. Another reason was that they were used to getting free water from the stand posts.

Hence, we focused on building trust in and the credibility of the public water supply system by explaining the need for reducing wastage and the redundancy of stand posts in the residential area when all households are provided with stable round-the-clock water supply. As the water supply system stabilized in a few months, public confidence and trust significantly improved.

Encashing this improved trust and confidence, the Jal Sathis were deployed to create a positive mindset in the localities in favour of the phased removal of stand posts.

Out of the 1,110 stand posts, all the 725 stand posts in the residential areas were successfully removed, and the remaining street taps in the high-footfall areas were replaced with new water faucets for the city's floating population. After three months of the trial run, incidences of breakdowns and faults became minimal and the team's confidence rose significantly.

Once we felt that things were fairly stabilized, we decided to go ahead with the formal inaugural event. I contacted Mr Pandian for a date when the CM would be available. The grand inaugural event was finalized for July 26 2021, and we started preparing for it.

When I started discussing the tentative plan for the inaugural event with Mr Pandian, my proposal was to keep it low-key. He said, 'What we have achieved is phenomenal.

DFT in our country is magical. We need to reach out to the outside world to convey this story. Let us go for media outreach at the national level so that this achievement inspires the rest of the country.'

Our team knew that we had achieved something great, but when it was validated by the PS to the CM, I was emboldened. We were advised to get the launch of the 24×7 DFT widely covered in local and national media including releasing a full-page advertisement in a leading English daily.

At 11 a.m. on that day, the CM dedicated 'Sujal' or the DFT Mission to all of Puri, which covered 2.5 lakh residents, including 66,000 urban poor living in slums. With this, Puri became India's first city to provide a 24×7 water supply that adhered to IS 10500 quality standards to every household; this water was fit for consumption directly from the tap.

Thus, Puri joined the prestigious league of global cities like New York, London, Los Angeles, Singapore and Tokyo by providing safe, round-the-clock water that is directly drinkable from the tap.

On this momentous occasion, the CM said, 'It's a landmark day, and a new chapter has been added to the development history of Odisha. All families in the Lord's abode will receive clean and hygienic drinking water directly from the taps from today onwards. Puri residents, tourists and pilgrims can now drink water from taps across the city, whether it be at home or the drinking water fountains. It has been my dream to provide piped water to every household in Odisha, and this is now turning into a reality.'

The inaugural ceremony was attended by several dignitaries, including Secretary Durga Shanker Mishra, IAS, of MoHUA, Government of India, who applauded the DFT Mission as the first of its kind and a replicable model for the whole country.

He said, 'It is a privilege to participate in an event where the CM of Odisha declared the ancient holy city of Puri as India's first city to have the distinct honour of 100 per cent tap connections with 24×7 water supply in every household and the facility of *DRINK FROM TAP*!'

The transformative move was beneficial not only for the residents of India's spiritual capital but also for two crore annual tourists and pilgrims. A total of 120 drinking water faucets were erected in public places along Grand Road leading to Jagannath temple; pilgrims would get 24×7 drinking water from these fountains.

This exceptional move would reduce the state's carbon footprint because, with access to drinking water around the city, pilgrims and tourists would no longer have to carry plastic water bottles. An estimated 400 metric tonnes of plastic waste generated by 3 crore plastic bottles could be avoided annually thanks to the mission.

The inauguration of DFT in Puri became a sensation across the country, and Puri was highlighted by national and international media. The MoHUA was also excited about this success in Odisha.

On the very next day after the inauguration, that is, on 27 July 2021, one senior official of SE rank from the New Delhi Municipal Corporation (NDMC) was sent to Puri by the MoHUA secretary to understand the nuances of the historical achievement.

Puri started receiving a host of visitors from several states and media houses across the country every day to understand the project. A 14-member expert committee of the Central Public Health and Environmental Engineering Organization (CPHEEO) constituted by the MoHUA for the revision of the Manual on Water Supply and Treatment and led by Dr M. Dhinadhayalan, adviser to the CPHEEO, also visited Puri on 21 and 22 October 2021. The committee was highly impressed by the world-class system of 24×7 water supply created in Puri.

Though the implementation of 24×7 water supply systems has been promoted by the Government of India since the 1990s and water supply systems across the country are generally designed for 24×7 supply following the CPHEEO manual, no city or state has successfully implemented the goal at a city level except pilot-level implementation in Hubli-Dharwad, Karnataka, and in Malkapur Nagar Panchayat, Maharashtra, about 10 years ago.

Inspired by the successful implementation of the DFT Mission in Puri, the Government of India got Puri's DFT story documented as a case study by the CPHEEO.

Our established practice is to do pilots only for the sake of learning while doing and developing an implementation framework in the form of SOPs and templates to facilitate scaling up across all cities without exception. The same approach was adopted for DFT as well.

On the day of the inaugural event, CM Naveen Patnaik announced the commencement of DFT Mission work in 17 other towns, including Bhubaneswar, Cuttack, Berhampur, and 14 OMBADC cities.

Seeing our success in Puri and our confidence in scaling up to 17 other cities, the Government of India mandated the implementation of 24×7 DFT quality water supply in all the 500 AMRUT cities across the country and earmarked a minimum of 20 per cent funds allocated under AMRUT 2.0.[24]

The experiences gained during the implementation of water supply projects under AMRUT 1.0 across the country, the inspiring successes of Odisha's universal coverage of tap

---

[24]'Atal Mission for Rejuvenation and Urban Transformation 2.0, Operational Guidelines', Ministry of Housing and Urban Affairs, Government of India, October 2021, https://amrut.mohua.gov.in/uploads/AMRUT_2.0_Operational_Guidelines.pdf.

water supply to every household and Puri's DFT Mission contributed to the idea of forming a national-level association of water supply utilities to create a platform to facilitate the exchange of knowledge and experiences.

The Association of Water Supply and Sanitation Organization (AWSSAR) is the first national association to proactively coordinate, promote and represent the collective interests of urban water and sanitation managers. It's a co-creation platform for knowledge-sharing among public sector water supply and sanitation organizations, parastatal agencies and ULBs engaged in delivering water supply and sanitation.

Odisha's impressive success stories of water supply, including the recent 24×7 DFT Mission in Puri, made the state the natural choice for the Government of India to organize the AWSSAR launch event. The MoHUA also wanted to give exposure to all state urban secretaries and heads of water utilities regarding the DFT Mission in Puri.

On 27 November 2021, Odisha hosted the launch ceremony of AWSSAR at Bhubaneswar. More than 60 senior officers from several states and union territories participated in the event. MoHUA secretary Durga Shanker Mishra, IAS, became the president of the association, and I, the principal secretary of the H&UDD of the Government of Odisha, was unanimously elected as its vice president-cum-CEO on this occasion. A case study of the 24×7 water supply project 'Drink from Tap Mission in Puri City' documented by the CPHEEO was also released by the Government of India so that all states could replicate the project.

Thus, Odisha has become the destination for learning best practices in urban water supply infrastructure and has started receiving policymakers and practitioners from several states and neighbouring countries regularly.

THE ESSENCE

The customer's perception is your reality.

—KATE ZABRISKIE

The success of the DFT Mission in Puri was driven by a people-centric approach that prioritized the needs and aspirations of the people. This approach required a significant shift in the officials' mindset and the adoption of the belief that 'consumer is king'.

Embracing this ideology led to behavioural changes in the officials, thereby building trust and people's confidence in the machinery. With the newly found consumer-friendly approach and consequent public trust and confidence, we received an encouraging response from the public, which fuelled our upscaling journey across the state. Placing the needs of the consumer at the centre of the initiative guaranteed success.

Consumer is king: bringing a cultural shift in water supply service delivery

14

SPREADING WINGS

After the successful implementation of the pilots and initiation of citywide upscaling in Puri, we started the planning exercise for DFT upscaling in other cities. I decided to tap into the funds available from the mineral-bearing districts, namely, the DMF and OMBADC, at the state level.

Drinking water is one of the top priorities under OMBADC and the DMF. We were getting funds from both for water supply infrastructure upgradation in some of the cities in the mineral-bearing areas. These were small projects proposed by the respective PHEO divisions to meet the existing gap in supply and demand rather than holistic development to achieve sustainable water security.

Knowing that OMBADC had huge funds, we proposed funding support for the universal coverage of pipe water supply in 14 cities under the OMBADC area. On 27 December 2019, the OMBADC Board gave in-principle approval for overhauling the water supply systems in the 14 cities[25] to provide a safe and clean water supply to every house with a tentative allocation of ₹1,200 crore.

---

[25]Sundargarh district: Sundargarh, Rourkela, Biramitrapur and Rajgangpur; Keonjhar district: Keonjhar, Joda, Barbil, Champua and Anandapur; Mayurbhanj district: Baripada, Rairangpur, Udala and Karanjia; and Jajpur District: Vyasanagar.

By this time, we had made substantial progress in achieving universal coverage of tap water supply across the state. This approval was partly in recognition of our performance and confidence in our team. Thus far, this was the highest single-source allocation in one shot for urban water supply in Odisha and it gave a big boost to our water supply agenda. Thereafter, we started preparing DPRs for each city.

Meanwhile, the pilot implementation of DFT was successfully completed and citywide upscaling had begun in Puri. As we were keen to upscale DFT in other cities, we decided to go for 24×7 DFT in the 14 OMBADC cities; accordingly, we made the DPRs for 24×7 supply and floated the tenders.

In the subsequent review meetings of OMBADC, I mentioned that based on the success of 24×7 DFT in the pilot areas, we were going ahead with the implementation of 24×7 DFT in the 14 cities.

One evening, I got a phone call from the then chief secretary who mentioned that the Oversight Authority appointed by the Supreme Court to oversee the functioning of OMBADC had spoken to him and raised concerns about the proposed 24×7 water supply projects in the 14 cities.

Upon receiving this information, I called up the Justice (Oversight Authority) to understand the concerns. Honourable Justice Mr A.K. Patnaik has always been supportive of people-centric development projects, and he was very keen to expedite the drinking water projects in the mineral-bearing districts.

During our conversation, he mentioned that he got to know that in India 24×7 is still in the experimentation phase and that there are several apprehensions like source sustainability, technical feasibility, high leakages and lack of expertise within the state. I told him about our success in the

12 pilot zones of two cities and the encouraging experiences in the ongoing implementation in Puri.

At the end of the conversation, I requested him for some time to make a presentation on the implementation strategy addressing the above apprehensions. In the meantime, we received a strongly worded letter from the office of the Oversight Authority on 13 October 2020, raising serious concerns regarding the implementation of 24×7 DFT in the 14 OMBADC cities.

The unexpected letter came as a big jolt and completely shattered our morale. I thought that my explanation would have convinced the Oversight Authority and we would be given the opportunity to make a formal presentation about our implementation strategy.

Though we should have put our plans on hold till a final decision was taken, I was determined to move ahead because basic infrastructure upgradation was never in question. The issue was only about the DFT component, which served as a top-up once the basic infrastructure was in place.

Holding back the project would have inordinately delayed the execution of the basic infrastructure component too as the tender process was already completed and Cabinet approval had been obtained. We were confident about what we were doing and went ahead hoping that our fieldwork would be able to convince OMBADC.

For various reasons, it took us almost 10 months to get a date to present our plan before the Oversight Authority and I was given the opportunity at the OMBADC board meeting on 6 September 2021. By that time, the Puri DFT had already been inaugurated and was a big national and international sensation.

Backed by our success and beaming with confidence, I made a strong case in favour of going ahead with our proposed 24×7 DFT plans. Though all their apprehensions were appropriately addressed in the presentation, our overwhelming success in Puri had tilted the case in our favour.

Finally, OMBADC was kind enough to approve funding for basic infrastructure upgradation in all 14 cities and for DFT components in seven major cities. It was agreed that the automation part of the project, which deals with the DFT component (around 5 per cent of the city-wise project cost), for seven cities would be derived from other sources. It was a winning moment for the DFT Mission and gave us big relief. The strength of the cause (supplying DFT-quality water 24×7), our passion and commitment could convince the OMBADC Board and the Oversight Authority.

The final sanction given by the OMBADC Board restored our confidence and we felt emboldened in our mission. In fact, OMBADC's sanction was key to upscale 24×7 DFT in 14 cities of Odisha, and the OMBADC became the country's first angel investor for 24×7 at such a large scale much before AMRUT 2.0.

We were marching ahead with implementation in all 14 cities with good momentum, and we were confident that we would show our gratitude and bring pride to the OMBADC by not only completing the projects on time but also surpassing their expectations in terms of qualitative implementation and public satisfaction.

Providing 24×7 DFT in mineral-bearing, underdeveloped districts was going to set a new benchmark amongst mineral-bearing areas across the country. This project is expected to benefit 1.6 million people living in the 14 cities, including the 0.4 million urban poor.

Cuttack is Odisha's second-largest city and is its millennium city and heritage city as well. It was the old capital of Odisha and it remains prominent with the high court and state police headquarters being located there. Though the people of Cuttack were proud of their city, they had begun nurturing the feeling that Bhubaneswar was getting all the importance and priority at Cuttack's expense.

To add to their complaints, only Bhubaneswar and Rourkela were included in the SCM. After making history in Puri city and deciding to upscale DFT in 14 cities with OMBADC funding, CM Naveen Patnaik desired Cuttack to be our priority.

Cuttack is a peculiar city surrounded by two major rivers: Mahanadi and Kathajodi. Though the city is surrounded by plenty of perennially flowing surface water, the drinking water supply system was dependent on groundwater with more than 225 production wells. To improve the water supply system in Cuttack, large-scale infrastructure upgradation work has been taken up by leveraging funds from the CSR, AMRUT, BASUDHA Mission and other sources.

As part of the state's policy, we started working on the migration of supply sources from groundwater to river water on priority. Accordingly, construction of the intake well, pumping system, WTP and transmission and distribution networks was taken up.

While implementing this agenda, we decided to take up universal coverage of tap water supply to every household under BASUDHA as our next agenda. When the CM decided to include Cuttack in the DFT Mission after Puri, we felt that the ground was ready to move to a higher goal. Cuttack has inherent challenges in terms of difficult field conditions such as unstable soil, high water table, dense population and narrow streets.

Because of the difficult field conditions, the sewer network project taken up with Japan International Cooperation Agency's (JICA) assistance[26] was prolonged for eight years,

---

[26]The Odisha Integrated Sanitation Improvement Project (OISIP) is supported by JICA at a project cost of approximately ₹4,000 crore and

hence causing inconvenience to the people. The unending implementation of repeated road cutting caused extreme frustration among the people of Cuttack.

In an old and densely populated city like Cuttack, implementing an underground sewer network project with the large-scale digging of roads and prolonged construction time can be compared with an open heart surgery on an aged, ailing person.

I remember that sometime ago I received a letter from a person in June informing me that his daughter's marriage was fixed to be held in December and asking me whether the road that had been dug up in front of his house would be repaired by the time of his daughter's marriage.

This letter made me realize how much inconvenience was being faced by the people of Cuttack and their anxiety while navigating the densely populated, narrow lanes where the roads were cut and left open indefinitely.

Keeping all this in mind, we worried that proposing another cycle of road cutting for laying the pipeline under DFT would result in public resentment. To our surprise, DFT's popularity in Puri had created a popular demand for it across the state.

When the CM announced that Cuttack would be the next DFT city, the city celebrated this announcement and felt rewarded after a long period of neglect. This created public support, which made our implementation journey smooth.

The successful implementation of an impactful programme can create huge public goodwill, which can significantly neutralize all previous negative sentiments. Preparing the people is extremely crucial before implementing any large-scale infrastructure project because during implementation,

---

works towards the augmentation of sewer coverage and sewerage treatment facilities in Cuttack and parts of Bhubaneswar along with improved drainage facilities in Cuttack.

public inconvenience is inevitable. In Cuttack, this aspect was missing in the implementation of the underground sewerage system with JICA's assistance, wherein the project was implemented without taking people into confidence.

As part of our implementation strategy, we decided to begin the project with three pilot zones in Cuttack. The High Court of Odisha is in Cuttack, so the judges, too, undergo the same ordeals and inconveniences faced by the people every day during the implementation of underground sewerage and drainage systems.

As a result, numerous public interest litigations (PILs) are actively pursued in the high court on various civic issues, including the underground sewerage and drainage systems. We realized that the judiciary's support was critical for the successful implementation of any project in Cuttack. The successful implementation of DFT had the potential to create goodwill, which could be relied on to deal with hurdles in the implementation of other infrastructure projects.

Hence, to demonstrate a highly impactful people-centric program like DFT and enhance the credibility of the city administration, we chose Judge's Colony (Ward No. 15) as one of the three pilot zones in Cuttack.

The other two pilot zones were Ward Nos 48 and 49 in the Jagatpur area with predominantly slum households. The pilots were successfully completed in a record time of four months, benefitting 24,000 people. The DFT for the entire city of Cuttack was then launched by the CM on 6 January 2022.

As I write this, DFT works for the entire city of Cuttack are in full swing and are slated to be completed by June 2024. From our past DFT experience in Puri for 2.5 lakh people, we have moved to a city with nearly 8 lakh people, which is a substantive jump in our upscaling journey.

Our next strategic decision was to implement DFT in

Berhampur, Odisha's third-largest city with 5 lakh people. I have already mentioned the significance of Berhampur as the largest city in southern Odisha and as an important commercial hub. With the completion of the GBWSP (Janibili), the city was ready for the next phase of transformation. Again, as part of our strategy, two pilot zones were chosen, and DFT was completed by the end of January 2022, thus benefitting a population of 15,000.

Upon the successful completion of the pilot, citywide implementation of DFT started. Unlike other cities, DFT implementation in Berhampur has been a quiet affair due to the model code of conduct imposed during the local body's election held between January and March 2022.

Though there was local demand for DFT services and the potential for a citywide upscaling ceremony with the CM, we proceeded with the implementation without the ceremony to avoid delays caused by the model code of conduct. The work is on the verge of completion (99 per cent) now.

After completion in Puri and commencing citywide implementation of DFT in Cuttack and Berhampur, I faced questions from several prominent people, including senior officials, such as: When are we getting DFT water in all of Bhubaneswar? Because DFT started in Bhubaneswar with the sub-pilot in Ishaneswar Slum followed by eight pilot zones, we made up our mind to commence citywide implementation soon after in some other cities.

Meanwhile, strategic planning for citywide implementation was worked out considering the inherent challenges of the capital city that is home to over a million people. Bhubaneswar, with a population of 11 lakh spread over 235 square km, has several challenges, such as being a temple city with over 100 heritage-protected monument sites, undulating terrain, old congested city areas, cosmopolitan IT hub and a fast-growing urban sprawl.

In contrast to other DFT cities like Puri with 19 district metering areas (DMAs), Berhampur with 40 DMAs and Cuttack with 56 DMAs, Bhubaneswar has the highest number at 108 DMAs. The implementation of the citywide DFT was taken up strategically by covering the city zone by zone.

As of now, including the initial pilot zones, 58 DMAs have been covered and the citywide work is going on in a phased manner targeted to be completed by March 2024. In October 2020, Bhubaneswar became the first million-plus city in the country to achieve 100 per cent HCs. In March 2023, Bhubaneswar became the first city to have 100 per cent metering of connections. Now, we are well poised to make Bhubaneswar proud again by making it the first million-plus city in the country to have 24×7 DFT.

The DFT journey that started in August 2019 with the sub-pilot in Ishaneswar Slum with 2,000 people moved to 1.85 lakh people in 12 pilot zones in two cities and then to 2.5 lakh people, thereby covering the entire city of Puri.

This was followed by the second- and third-largest cities of Odisha with populations of 8 lakh and 5 lakh respectively, to the capital city with 1.1 million people. The mission has now moved to cover a total of 24 cities with a cumulative population of 2.5 million.

The entire journey of supplying 24×7 DFT-quality water in Puri and the pilot zones of Bhubaneswar, Cuttack, Berhampur, etc., attracted global attention too. As a result, WATCO received the prestigious Distinction of Global Water Leader Award 2022, at the Global Water Summit on 17 May 2022 in Madrid, Spain. The Global Award was presented to Odisha for achievements in providing 24×7 DFT-quality water in the state and accomplishing 100 per cent HCs in the urban areas under WATCO.

WATCO was shortlisted for the award with three other agencies: Beijing Drainage Group, China; Lusaka Water

Supply and Sanitation Company, Zambia; and Sedapar, Peru. Global Water Intelligence (GWI) announced the awards at the Global Water Summit 2022 award ceremony, which recognized the industry's greatest achievements in 2021.

The award was received by me as the chairperson of WATCO on behalf of the Government of Odisha. During the award ceremony, the winners were announced under various categories. The public water utility category was announced first. WATCO had applied under this category and was shortlisted as one of the top four. There was pin-drop silence in the audience, and everyone was sitting with bated breath.

Though I was trying to control my anxiety, my heart was racing and I was ecstatic when WATCO's name was announced. When I went to the dais and received the award, I remembered two people. One was CM Naveen Patnaik without whom I would not have been the principal secretary of the H&UDD for eight long years and without whose blessings WATCO would not have become the Global Water Leader within just five years of its inception. The other person I remembered was my mother who was a schoolteacher and to whom I am forever indebted for the love she gave me till her death.

We continued our upscaling journey with more vigour and passion. At this juncture, we consciously decided to complete and commission 24×7 DFT projects in at least one pilot zone in each city instead of waiting for the completion of the mission in the entire city.

This would yield two benefits. First, we would be able to demonstrate the results, win public trust and confidence and gain public cooperation to complete the project in the city in a timely manner. Second, the WATCO engineers would get on-the-job training, which would boost their confidence for citywide execution. The intention was that it would demonstrate the benefits quickly, which would help create public confidence in DFT work and the WATCO team would get trained in end-to-end DFT execution.

Another momentous occasion and milestone in our journey was on 21 December 2022, when CM Naveen Patnaik inaugurated the 24×7 DFT in the 58 zones of 19 cities to the public and declared Gopalpur the second city in India to have complete coverage under the DFT facility after Puri. Subsequently, in our upscaling journey, another five cities with complete coverage of DFT benefiting 0.2 million population along with partial completion in 13 other cities benefiting 1.1 million was inaugurated by CM Naveen Patnaik on 9 October 2023. As of December 2023, more than 2.5 million people in 24 cities of Odisha are getting 24×7 'Drink from Tap' water in their homes and the WATCO team is very much on track to cover all the 115 cities of the state within the next five years.

THE ESSENCE

Teamwork is the ability to work together toward a common vision. The ability to direct individual accomplishments toward organizational objectives. It is the fuel that allows common people to attain uncommon results.

—ANDREW CARNEGIE

When a group of people with a shared vision and passion work together, the results can be miraculous. The power of a dedicated team is magical.

Odisha's DFT is a testament to what can be achieved

when a dedicated team works with passion and the shared vision of strong political leadership.

Team building and perseverance are the keys to successful project implementation!

15

CONFRONTING CHALLENGES, CRAFTING SOLUTIONS

While implementing the DFT Mission, we encountered innumerable field challenges at different stages. There was no guidebook readily available to refer to for a quick solution, so we had to devise innovative strategies to overcome the challenges and move ahead.

Every time we hit a problem, we focused on the possible solutions rather than worrying about the issues. The solutions emerged through continuous brainstorming sessions amongst the team, which were small victories overcoming each obstacle.

Each obstacle and its solution not only gave us tremendous hands-on learning but also trained us to find solutions every time we hit a roadblock. Here are some of the challenges we encountered during implementation and how we successfully sailed through them.

1. NON-AVAILABILITY OF EXISTING INFRASTRUCTURE ASSETS INFORMATION

Challenge

When we embarked upon planning for DFT, we had to find out the current status of existing infrastructure laid at different points in time and under different projects. Getting reliable data regarding such assets is always difficult because record-keeping is often not well organized in water supply offices.

Generally, record-keeping is poor in engineering departments, and the incumbents often get transferred or superannuated, resulting in the loss of institutional memory.

Solution

This issue was largely managed through various ways, including finding the helpers, pump operators and linemen who had retired from service (even as long as a decade ago) to get information about pipe alignment and other asset details.

Consequently, we could get more accurate information about the asset details than was available in office records. Additionally, interactions with local communities, especially aged people, were also very helpful. With thorough field surveys, we mapped each asset on the GIS platform to identify the infrastructure gaps to be filled for the DFT.

We realized that the records available in the PHEO office provided only incomplete, piecemeal information. Our interactions with the old grassroots-level field employees, community engagements and thorough field surveys brought authenticity and completeness to the infrastructure asset data.

2. NON-AVAILABILITY OF CONSUMER INFORMATION

Challenge

Unlike the banking sector wherein Know Your Customer (KYC) is an established practice and the customer profile is used to provide better services, the same is not the case in the utility sector, including water supply.

Details about HCs, consumption patterns, payment details, etc., were not readily available. There was a total absence of consumer connect to the extent that the water supply officials did not know their consumers and the consumers hardly knew the water supply authorities.

Solution

The essence of DFT is the quality of water supplied to every home; therefore, it becomes important to know the consumer, establish a consumer connect and improve the quality of service. Hence, we felt that mapping the location and details of each consumer was essential for planning, designing, establishing and running a DFT system that would meet the customer's expectations.

Door-to-door consumer surveys and developing a GIS-based database of consumers with all the required details helped collect actual field information with geographical locations and facilitated accurate planning and smooth implementation. Thus, we were ready with the information needed about our consumers (existing and potential) for universal coverage and improved services.

3. RESISTANCE TO 100 PER CENT HOUSE CONNECTIONS AND METERING

Challenge

Taking an HC was not acceptable to everyone, especially to two kinds of people. First, those who already had household borewells built at their own cost, which could be operated at will with the owners not having to pay any recurring water charges.

The second category consisted of those who managed to take illegal connections and enjoyed free water. Additionally, a few people engaged in the water supply business feared that their businesses would be affected if everyone took water supply connections.

Also, when we made it clear that as part of DFT every HC would be metered, many feared that metering would result in increased monthly water charges; hence, they opposed metering.

Solution

The resistance to HCs was tackled by creating public awareness and community engagement through Jal Sathis. In addition, policy measures such as HCs being executed by water supply authorities as public work, introduction of an instalment system for HC payment, making HCs free for the urban poor, exemption from road-cutting charges and a hassle-free connection process contributed towards encouraging the public to take HCs, which helped us achieve 100 per cent HCs without leaving anyone behind.

We faced more challenges during the installation of water meters. People were used to paying fixed minimal charges with unlimited consumption. Hence, the concept of metering was not well received. They realized that with metering they would have to pay for their actual consumption.

Contrary to our assumption, it was relatively easy for us to convince the slum dwellers about metering and payment on the basis of actual consumption. Instead, there was more resistance in the posh areas when dealing with the urban elite. We realized that consumption in urban poor settlements is not only low but also linked to need, whereas in the higher-income areas, the unmetered water supply was used for various purposes, including gardening and car washing, with unchecked wastage.

Hence, the elite resisted the installation of meters. Intensive efforts were made to sensitize the households and build local pressure through Jal Sathis and local community leaders.

The fact that metering was made non-negotiable and all households, including slums, were covered under metering without any exemption made everybody accept the inevitable. In some places, 100 per cent metering in slums was cited as an example to crack the hard nuts.

4. HOW TO ENSURE THE 'DRINK FROM TAP' (IS 10500) QUALITY OF WATER AT EVERY CONSUMER TAP

Challenge

It is easy to control quality parameters at a few centralized production and treatment points. The real challenge is to ensure DFT quality in the tens of thousands of consumer taps after the water travels through several kilometres of underground pipes from the treatment point by criss-crossing several drainage and sewer lines en route.

Even if we ensure that the desired quality reaches consumers' homes, the same quality cannot be maintained when the water is stored in their overhead tanks or underground sumps, which are prone to contamination and we have hardly any control over them.

Solution

I am often asked about the treatment process and technology adopted for providing DFT-quality water. I clarify to them that the secret is not in the process or technology but in ensuring integrity in leak-proof pipe networks and the deployment of smart water management to monitor the critical parameters.

Real-time residual chlorine analysers at the far end of the network along with automatic chlorine-dosing systems have been installed to ensure safe water even at the farthest consumer end.

As a part of our water quality monitoring and surveillance protocol, state-of-the-art PPP water testing laboratories have been set up at regional levels for the regular testing of all required parameters from strategic sampling points. Additionally, mobile labs on wheels are deployed for on-the-spot water quality testing.

People were sensitized about potential water contamination in their overhead tanks and underground sumps. They were made to realize that unless the water is directly taken to the taps by bypassing intermittent storage, contamination cannot be prevented. Maintaining adequate water pressure and 24×7 availability demonstrated that household storage tanks are now redundant.

In spite of awareness creation, households wait and watch for a few months before finally abandoning the household storage tanks once trust and confidence in the consistency and reliability of supply develop.

5. IN-HOUSE KNOWLEDGE GAP AND LACK OF EXECUTANTS WITH EXPERIENCE IN 24×7

Challenge

The state's lack of prior experience in implementing 24×7, a lack of sufficient precedence even across the country and the unavailability of an implementation manual were our main challenges. Knowledge and prior experience gaps existed at all levels, from top management to field-level plumbers.

Solution

To start with, we sent a core team of key WATCO engineers to undergo a four-day training programme at ASCI, Hyderabad. Professor V. Srinivasa Chary, director of ASCI, Hyderabad, has been training water utility engineers and championing the cause of 24×7 in the country for two decades. The team was inspired by his input during the training programme and was motivated by his passion for 24×7. Upon their return, I met the team and took their feedback.

I was happy to hear Mr P.K. Swain say, 'Sir, we can do it,' with confidence in his voice, which made me make the next moves faster. We started exploring further sources of learning from a successfully operating 24×7 system. We got to know

Professor Asit Biswas, a native of Odisha, renowned global water expert and recipient of the Stockholm Water Prize.

Through him, an almost customized one-week training programme was organized and six of our key members were sent to the SWA of the PUB, Singapore. The team was exposed to the nuances of the high-quality 24×7 water supply system that had been operating in Singapore for decades.

Though the team learnt the basics of 24×7 water supply technology, including the design aspects, from Professor V. Srinivasa Chary, the Singapore training exposed them to live operations. The team learnt operational aspects and other equally critical components of 24×7 such as consumer connect, customer relations, quality assurance, smart water management and water loss management.

Upon returning, this team laid out the roadmap for the DFT 24×7 water supply system. They started implementing the programme and training the other engineers and staff of WATCO. In this way, we were able to significantly bridge the knowledge gap.

When we started preparing the tender document to engage executing agencies, we realized that we would not get executing agencies with experience in the 24×7 framework. In the absence of such experienced executants, we decided to engage regular water supply contractors and have our core team train them on the job regarding the 24×7 system.

This was not a one-time activity. Engaging regular water supply contractors and training them on the job is still continuing. This is done to ensure adequate participation of bidders in the tendering process to avoid single tenders and facilitate competitive bidding. Thus, we made the tendering aspect successful, which is critical in the upscaling process. Moreover, a larger pool of experienced workers in the private sector is also being created.

We often focus on the larger aspects and ignore the

smaller but critical ones. Even a state-of-the-art 24×7 water supply system designed and built by eminent engineers may fail due to leaking HCs because of poor workmanship on the ground. Training plumbers is one such small thing that we did not lose sight of. Local plumbers were trained and empanelled with WATCO for the execution of leakproof HCs.

By this time, we needed to institutionalize capacity-building activities in a continuous and sustainable manner. Taking cues from the SWA, which is the capacity-building and research institution of the PUB, Singapore, we decided to establish a similar institution in Odisha.

We established the Odisha Water Academy (OWA)[27] with the aim of it functioning as a 'centre of excellence' for creating a trained workforce and continuous capacity building of staff. The OWA imparts training and capacity building in the water supply and sanitation sectors for all levels from operators to policymakers.

Though we created the OWA to meet our own training requirements, it has now become an institution of national repute that trains more than 3,000 people, including secretaries of urban development departments, chief engineers, private contractors, consultants from within India and also from neighbouring countries. The OWA has been empanelled as a premier training institute by MoHUA.

The OWA has now partnered with institutions of national repute like the National Institute of Urban Affairs (NIUA) and ASCI to provide national- and international-level training programmes.

[27] The OWA was established under WATCO of the H&UDD, Government of Odisha. It is committed to offering high-quality practitioner-focused training solutions and exposure visits for diverse stakeholders in the water and wastewater sector through self-financing.

6. HUMAN RESOURCE CONSTRAINTS IN CONSUMER SERVICE DELIVERY

Challenge

Between 2015 and now, we moved from 40 per cent network coverage to more than 95 per cent network coverage and from 3 lakh HCs to 1 million HCs. The scope of activities has grown manyfold with the corresponding human resource requirement. A corresponding increase in staff strength through large-scale government staff recruitment is extremely difficult.

At the same time, we needed additional human resources to meet customer service delivery standards without any time lag. With limited human resources, timely meter readings, bill delivery and revenue collection, which are essential to maintain the standard of services, have become our challenges.

Solution

Odisha has had a strong women's SHG movement for the past two decades. We have actively engaged their services under various programmes of the H&UDD, Government of Odisha.

Along with the Department of Mission Shakti, ULBs also form women SHGs. Initially, they were engaged in income-generating activities such as tailoring and making snacks, papads and pickles, which involve lower-level skill sets.

A few years ago, the department embarked upon a journey of women empowerment by partnering with them in operating Aahaar[28] centres under the affordable meal program and operating and managing septage and SWM facilities that require financial and techno-managerial skill sets.

---

[28]Similar to affordable food provision programmes across the country, Aahaar was launched in Odisha in 2015. With the assistance of the State's MSG, the urban poor are provided with food every day at the cost of only 5 rupees.

Having experienced the benefits of partnering with community-based institutions like women SHGs, I was inclined to partner with them in the water supply sector too. However, many people believe that with water supply being a highly technical subject, entrusting the responsibility to women SHGs may be disastrous.

I was fully convinced that involving women SHGs at the grassroots levels would not only address my human resource shortages but also bring various other benefits. Detecting and regularizing illegal and unauthorized connections, mobilizing new connections for achieving 100 per cent HCs and revenue collection at consumers' doorstep were services that were difficult to accomplish with regular government staff.

Also, the consumer connect that was absent earlier could be achieved only by the women SHG members who were chosen from the local community itself. Therefore, I considered community partnership as the solution to several problems. Hence, we engaged members from local women SHGs such as Jal Sathis on a performance-based incentive system.

Jal Sathis act as a bridge between WATCO and the consumers. They are trained and entrusted with meter reading, bill distribution and doorstep revenue collection with MPoS machines. They were also provided with a water testing kit and trained to carry out household-level water quality indicator tests. There is a specific amount for carrying out each of the above activities (as mentioned in Chapter 8), and a 5 per cent incentive for revenue collection.

Our model of leveraging the power of community partnership with Jal Sathis as change agents has transformed water supply service delivery and taken women's empowerment to new heights. With this smart move, we created a large workforce within a short span of time for improved service delivery without going for large-scale government staff recruitment.

Jal Sathis played a crucial role in urban Odisha achieving 100 per cent piped water HCs in 105 out of 115 cities so far.

Revenue collection efficiency has improved substantially. For example, in Puri, it increased to 97 per cent. The community partnership model of engaging women for water service delivery has received several accolades at the national level.

Inspired by the success of community partnership in Odisha's water sector, MoHUA made community partnership an essential ingredient in water supply service delivery for the national flagship programme, AMRUT 2.0, making other states emulate the Odisha model. With this smart innovation, Jal Sathi has become a game changer in the success of the 24×7 DFT water supply delivery service in Odisha.

7. PUBLIC TRUST AND CONFIDENCE

Challenge

Public water supply suffers from a legacy of poor water quality and lack of public trust. Generally, we do not trust the quality of water supplied by public authorities. Instead, we trust water procured from private vendors; even an unbranded water pouch garners more trust than the water supplied by public authorities.

When I asked my chief engineer heading the water supply division to drink water directly from the tap, he was reluctant because he did not have confidence in the quality of water supplied by him. He preferred a mineral water bottle.

I realized that if this is how much we trust our water, how can we expect the common person to trust the quality of our water? Further, the public was apprehensive about whether the government would be able to maintain a continuous water supply without break and with consistent pressure.

The elimination of overhead tanks and sumps was essential to prevent water contamination, and consumers were advised to bypass their household overhead tank and sump and use water directly. As the public had already invested in and built overhead tanks and sumps, they did not want to abandon them by believing the water supply authorities but preferred to wait and watch.

Solution

Seeing is believing, so we decided to display key water quality parameters on a real-time basis on large LED display boards at high-footfall locations across the cities.

Apart from extensive IEC campaigns (Pure for Sure campaign), measures like on-the-spot water quality testing by mobile labs on wheels and random quality testing of water in consumer taps by Jal Sathis gradually help build public trust and confidence.

8. NON-REVENUE WATER

Challenge

Non-revenue water (NRW) is the water produced that does not earn any revenue. It comprises (a) consumption that is authorized but not billed, such as PSPs; (b) apparent losses, such as illegal water connections, water theft and metering inaccuracies; and (c) real losses such as leakages in the transmission and distribution networks.

When we started our journey in 2015, as per a study conducted by an independent agency, Odisha's NRW stood at around 54 per cent. This meant that 54 per cent of the water produced was either getting lost on the way or not being billed. As per baseline data from Puri, NRW was 47 per cent. We understood early on that unless we controlled

NRW or water losses, we would not be able to sustain 24×7 water supply services.

One of the major risks in providing 24×7 water supply is that if there are leakages in the pipeline, it results in 24×7 water loss, unlike intermittent supply wherein the loss duration is limited to supply hours. One of the challenging aspects of dealing with leakage is to identify the point of leakage because most of the pipe network system is underground (mostly one metre below the surface). Water, by virtue of gravity, tends to flow downwards when there is a leakage.

At times, it takes several months for the leakage water to become visible above the surface. The quick identification of leaking points is critical to NRW reduction. This is one of the most challenging areas of 24×7 water supply for which new technologies are evolving worldwide.

Similarly, if water is supplied and not billed, 24×7 supply will encourage and result in irresponsible consumption. At the design stage, each component of the 24×7 system was designed taking into account the population to be served and the per capita supply to be ensured with consistent pressure. With large-scale leakages and irresponsible consumption due to non-billing, demand will become erratic, thereby making the supply system technically and financially unsustainable.

Solution

Reducing NRW has been given top priority right from the planning stage. Strategies were developed to address various aspects of NRW reduction and implemented in a mission mode with clearly defined timelines, milestones and measurable outcomes. A ferrule point is the T-joint connecting the street pipeline to the pipeline going to the house. These HC ferrule points were identified as potential leaking points due to unstable joints.

Every defective ferrule point acts as a point of leakage and contamination. In a street pipeline of about 1,000 metres, we may have a ferrule point at five-metre intervals, amounting to 200 potential leakage and contamination points to be tackled.

Due to the enormity of the challenge and especially because the USP of our DFT Mission was to provide drinkable (IS 10500) water at every tap, we decided to replace all the existing ferrule joints with new ones with saddle and compression fittings to make them leak-proof and contamination-proof. The saddle provides stability to the ferrule joint and the compression fitting makes it leakproof.

An exclusive NRW cell with a dedicated team of engineers was established at WATCO's head office to guide and monitor the implementation of NRW control measures in the cities. The NRW cell in every DFT city is equipped with state-of-the-art leak detection equipment, and the team was rigorously trained for a month to effectively carry out NRW reduction activities.

The industrial Internet of Things (IoT) system embedded with multiple pressure and flow monitoring sensors at strategic locations across the pipe network helps identify leakages in case of an abrupt loss of pressure or flow in the pipe network, which can be graphically seen online on a real-time basis using the smart water management system. This has helped with the quick identification and repair of leakages, helping control NRW substantially.

9. QUICK COMPLAINT REDRESSAL

Challenge

The manual mode of a consumer lodging a complaint and staff assignment for resolving the complaint used to take days to weeks, resulting in consumer dissatisfaction. Many times, identifying the problem itself took most of the time.

In the absence of consumer connect, consumers generally preferred not to register complaints, partly because the complaint lodging process was cumbersome and partly because they were not hopeful of a response from the authorities.

They preferred living with the problem of dissatisfaction. We needed to simplify and streamline the complaint lodging and redressal process to improve consumer satisfaction and operational efficiency in service delivery.

Solution

To simplify and streamline the process, an interactive voice response system (IVRS)-based centralized customer care system was established for quick complaint lodging, automatic assignment of the complaint to the concerned engineer/crew for immediate action and online, real-time tracking of resolution.

For immediate response to complaints, leakage control and supply-related incident management, quick response teams with exclusive mobile vans and a dedicated, skilled crew are deployed with the necessary tools and equipment. The quick response team is available 24×7 to rush to the incidence site for quick redressal. The deployment of these teams also helps reduce NRW by quickly repairing leakages.

The complaint lodging and redressal process along with the functioning of the quick response team is monitored in real time in the smart water management system. The prompt deployment of quick response teams to the incidence spot for complaint redressal has greatly enhanced customer satisfaction and worked as a confidence-building measure. Even if the quick response team is unable to immediately resolve the issue in case of major repairs, their immediate arrival to attend to the complaint comforts consumers.

10. PROJECT MONITORING

Challenge

When supply of 24×7 DFT-quality water was promised to the public, it was essential that the promise be delivered in a time-bound manner. As we had raised people's aspirations and significant hype is already created around DFT, any delay in completing the project would dent our credibility. Our self-set, ambitious timeline required the implementation of professional project management practices at all levels.

Solution

Speed and scale have always been our urban team's USP, so we were conscious of closely monitoring and arresting any slippage from targets. For real-time project progress monitoring, a web application on a project performance monitoring system (PPMS) was developed and deployed. Field functionaries update their item-wise progress data from the field in real-time. This helped in the quick identification and resolution of field challenges for timely project completion.

As part of the 5T initiative strategy, the fourth Saturday of every month was earmarked for reviewing progress under 5T initiatives. DFT is a 5T initiative, so we decided to hold the review meeting every Saturday. The Saturday meetings were institutionalized from the very beginning of DFT; they are held by the CEO of WATCO with key field officials.

Since the inception of DFT in August 2019, not a single Saturday has passed without a DFT review at the CEO level. Such is the passion with which the mission has been implemented and monitored, which I believe is one of the reasons for our success so far. Though Saturday meetings to review the 5T initiatives were prescribed by the government at the highest level, I can confidently say that this DFT team led by the CEO of WATCO ritually held the meetings sincerely and in a meaningful manner.

I salute Mr P.K. Swain, my field lieutenant, who is leading by example with passion and commitment.

THE ESSENCE

We should not give up and we should not allow the problem to defeat us.

—A.P.J. ABDUL KALAM

Did you know that over 2 billion people worldwide lack access to safe drinking water? That's more than a quarter of the world's population today! These staggering statistics highlight the urgent need for initiatives like the DFT Mission. When we created this mission, I knew it wouldn't be smooth sailing but riddled with innumerable obstacles and stumbling blocks en route.

Challenges and failures are inevitable in any pursuit, and it is often easy to give up by putting the blame on some obstacle. We were amply clear from the beginning that failure was not even a distant option and that the mission was designed only to succeed come what may. Thus, we were determined to face challenges head-on with the persistence to keep resolving each one of them and continue moving.

Strong leadership, effective project management practices and a commitment to innovation and problem-solving can help overcome even the most complex challenges and achieve successful outcomes. Knowing the strong leadership that will not accept a 'no' under any circumstances, the team started innovating to find solutions even for insurmountable challenges.

A few examples are finding retired staff, interacting with

local communities to establish missing infrastructure asset data and engaging with reluctant urbanites to ensure 100 per cent metering. With innovative strategies, the team demonstrated that nothing could stand in the way of achieving their goal of providing 24×7 DFT in Odisha.

Comprehensive solutions to the challenges: key to revolutionizing water supply in Odisha

16

HARVESTING DREAMS

Nothing is more satisfying than seeing disadvantaged communities benefiting from government schemes. Unlike other sections of society, these communities typically rely solely on government interventions to improve their quality of life.

A key challenge in developing countries like India is translating political vision into ground reality. However, today, 97 per cent of urban households in Odisha have piped water connections, with 105 out of 115 cities achieving 100 per cent HCs. Before June 2024, all households in urban Odisha will have piped water connections.

Moreover, out of around one million water supply connections across the state as of now, about 4,00,000 connections are in slums benefiting the urban poor. Water connections in slum households were earlier considered not permissible due to landownership issues until we announced a 'leaving no one behind—universal coverage' policy.

Before the drinking water transformation in Odisha, the poor sections of urban society had intermittent or no access to piped water, and were thus forced to consume water from other potentially contaminated sources. This resulted in many cases of waterborne diseases every year.

The lack of access to piped water supply at home disproportionately affected the women and children. They were compelled to spend hours fetching water when they could instead have been engaged in income-generating activities (for women) and attending school (for children).

This scenario has completely changed. Drinking water is now available in every household, thereby significantly reducing the scope for consumption of contaminated water and eliminating the incidence of public health outbreaks. Because water connections are available within household premises, children can improve their attendance in school and women can engage in economic activities and help augment their household income.

The findings of an independent impact study were very encouraging. During the ground survey of the impact study, 80 per cent of the slum respondents reported that before the universal coverage of water supply, they used to spend about an hour and a half fetching water.

Invariably, women and girl children took on this responsibility. With a tap water connection within their premises, these women and girl children can now utilize that time for other productive tasks. It was very encouraging to learn from the survey that most of the girl children are now using that time for educational activities.

The policy measures to treat HCs as public work, introducing an instalment system for the payment of connection charges and waiving connection charges for the poor have provided great relief to economically weak families.

To observe the real impact of our actions on the ground, we collected waterborne disease data from the IDSP under the Department of H&FW, Government of Odisha, between 2014 and 2021. I was truly amazed looking at the information. In 2014, there were 3,608 jaundice cases with 29 deaths, decreasing to 1,776 cases and 4 fatalities in 2015. Since 2019, urban areas have reported zero jaundice cases.

Similarly, diarrhoea cases have also dropped significantly, with less than 200 per year and no fatalities in urban areas since 2020. It is indeed satisfying to see that after safe water was provided to households, waterborne disease cases have dropped by more than 90 per cent with zero fatalities in the last 5 years. With these achievements on the ground, I clearly understood that 'Investment in water is nothing but investment in health.'

While the focus across the country was still on providing piped water supply, we set a goal of quality assurance and graduated to the DFT Mission to provide IS 10500-quality drinking water from the tap around the clock.

The commissioning of the 24×7 DFT Mission in Puri (the first of its kind in India) has brought about a change in people's expectations from water supply authorities across the country. Even within Odisha, people, especially the elected public representatives, started asking when DFT water would be provided in their cities.

THE 'DRINK FROM TAP' MISSION HAS LED TO SEVERAL TRANSFORMATIONS IN PURI

Puri has earned the distinction of being the first city in India to provide 24×7 DFT-quality water to its residents and joining the prestigious league of cities like New York, London, Tokyo and Singapore.

Further, it has reduced the stress on groundwater resources because there has been a significant drop in the number of people using hand pumps and borewells after the implementation of the DFT Mission. The women who were once victims of the system have now become Jal Sathis by partnering with the DFT Mission. Thirty-two members from women SHGs were inducted for water service delivery at consumers' doorsteps.

The water supply system is now modernized with large-scale automation, which was earlier managed manually and was error-prone. The mission has leveraged technology to deploy a smart water management system[29] that helps monitor water quality and track leakages in real-time, resulting in a steep decline in water loss from 47 per cent to 15 per cent.

Its impact on the reduction of plastic waste in Puri is substantial. The provision of water faucets across the city has facilitated round-the-clock access to safe and clean drinking water for the 20 million tourists visiting Puri annually, thereby eliminating the need to buy bottled water. This has helped potentially avoid plastic waste to the tune of 400 metric tonnes annually, considering each tourist uses at least 1 plastic water bottle of 1-litre capacity, which weighs about 20 gm.

Further, by ensuring 24×7 water supply (DFT), the need to construct household-level water storage tanks has been eliminated, helping families save about ₹50,000 per house. It has also led to savings in energy costs as household water treatment is no longer required. Family savings have also been safeguarded by eliminating the recurring cost of water cans for drinking and cooking, which was burning a hole in consumers' pockets.

I have been asked on several occasions, 'What has been the impact of the DFT Mission?' This was a very valid question because, at the end of the day it is natural to enquire whether the efforts, time and resources invested in this mission are justified.

In June 2020, I invited the NIUA to carry out an independent assessment of the pilot schemes of the DFT Mission that were implemented in Bhubaneswar and Puri. My brief to them was clear: tell me explicitly if something is amiss and what improvements are still needed so that we

[29]Smart water management includes real-time data collection, storage, retrieval, analysis, decision support, and command and control of public water supply systems using an array of IoT technologies.

can fix them when we scale up the DFT Mission to the remaining cities in the state.

The NIUA, a think-tank established under MoHUA, Government of India, had done significant work in the WATSAN sectors that made me entrust them with this mandate.

I had several discussions with the NIUA team to make them understand the nature, scope and objective of the evaluation. Once I was satisfied that the suggested evaluation framework would bring out the expected learnings from the ground on how the DFT Mission works in the field and how much it has benefited the people and impacted other key stakeholders, I stepped back and let them perform the assessment independently. I also instructed my colleagues at WATCO to not interfere with the evaluation but to provide all necessary data and logistics support to the team.

The evaluation framework had three broad categories of investigation. The first was system efficiency to understand whether the DFT system could eliminate the technical challenges typically associated with intermittent systems. The second was the consumers' perspective to get first-hand feedback from the people for whom the DFT Mission was implemented. The third was the sustainability and replicability of DFT. It aimed to investigate whether the current policies and systems were adequate to ensure that the DFT Mission could sustainably serve the people with the same outcomes in the future and if this model was replicable in its current form to other cities.

NIUA completed the evaluation in seven months and presented it to my WATCO colleagues and me. The study had several interesting findings. Most of the findings aligned with what we had imagined and planned from a system efficiency standpoint.

For example, NRW in Bhubaneshwar and Puri came down

to 15 per cent from a staggering 47 per cent before DFT. This is important because significant financial resources go into producing DFT-quality water. It would be criminal if this water were lost to leakages, theft and other inefficient practices.

We could bring down NRW to around 15 per cent, which is quite significant as the national service-level benchmark requires cities to bring down NRW to below 20 per cent. However, I am aware that progressive water utilities worldwide have reduced their NRW to below 10 per cent. This will be our target as we implement the DFT Mission in other cities.

Another vital aspect that emerged from the system efficiency analysis was 100 per cent metering in all DFT zones. While this was no surprise because we had planned for it, what was impressive was that more than 95 per cent of these meters were functional. Meter faults and non-functioning meters are general challenges in the metering process, so we made the implementing agency responsible for the procurement, installation and replacement (in case of defects within five years) as part of the contract.

This has become one of the strong points that freed us from the onus of establishing meter repair facilities and the associated pains of meter maintenance. It has not only helped monitor system performance closely but also enhanced bill collection efficiency to above 90 per cent in all DFT areas.

In an interesting analysis, the NIUA study[30] compared the DFT system in Puri to water supply systems in similar-sized cities across the globe, including Aarhus (Denmark), Brussels (Belgium), Canberra (Australia), Charleston (USA), Chittagong (Bangladesh), Christchurch (New Zealand), Grozny (Russia), Hanoi (Vietnam), Helsinki (Finland),

[30]The study titled 'Evaluating the Efficiency and Impact of Drink from Tap Schemes in Odisha' was carried out by NIUA in 2022.

Incheon (South Korea), Katowice (Poland), Kimberly (South Africa) and Wellington (New Zealand).

The results of the analysis were that Puri's DFT system matched the water supply systems in the other cities on almost all parameters and, on certain parameters, was better than others. For example, in terms of water supply coverage by population, bill collection efficiency and energy usage, Puri was amongst the best performers. Likewise, amongst all the cities surveyed, only Aarhus, Canberra, Charleston, Incheon and Katowice had better NRW rates than Puri.

The only parameter that Puri performed poorly compared to others was the cost recovery indicator. At 37 per cent, cost recovery in Puri was much lower than that of the others, with most cities boasting 100 per cent cost recovery.

Cost recovery has been on my mind for quite some time now. If all the cost has to be recovered from user fees, consumers in Puri will have to pay at least three times what they are paying today for one unit of water. This is difficult for the consumers to pay and the water supply authorities to collect.

I am, therefore, inclined to look at cost recovery from various perspectives. The first is the need to introduce a telescopic tariff system. Next, we must explore avenues of revenue augmentation other than user fees, such as renting out unused office space/land and raising revenue by introducing financial instruments like water bonds.

There is also the option of considering providing drinking water as an essential public service, similar to making and maintaining hospitals, roads, public parks, etc., and using taxpayers' money without resorting to full cost recovery from user charges. Therefore, an annual gap-filling budget will have to be provided to operate the water supply system while adhering to service-level benchmarks with reasonable cost recovery.

Providing good quality water is not only an essential public service but also an essential ingredient of preventive public health. It has more economic returns due to its contribution to the population's productivity, especially in developing countries where a substantial population lives on the fringes of poverty.

Hence, it would be inappropriate for me to look at public water supply from a very narrow perspective of financial returns on the investments made and target full-cost recovery for such an essential public service.

The findings regarding customers' perspective were a revelation. In just over nine months of DFT service, the surveyed households reported a dramatic drop in groundwater use. This figure was 60 per cent before DFT and came down to 2 per cent after DFT.

For me, this is a significant achievement because prior to DFT, it was challenging to regulate groundwater extraction due to the large number of users. Now, there is less unregulated use of groundwater, which bodes well for the sustainability of water resources.

In another finding that was personally satisfying for me, more than 80 per cent of the survey respondents reported that they were now drinking water directly from the tap with DFT connections at home. Now, this may seem a bit strange at first glance. After all, as the name suggests, DFT water is meant to be consumed directly from the tap. If that does not happen, the mission has failed. From a service provider perspective, merely providing the service does not necessarily mean that the services will be used. In addition to providing high-quality water, it was equally vital to earn consumers' trust.

Years of apathy on the part of city officials over the suffering of people due to a lack of access to drinking water had led to a loss of trust in government institutions. From the DFT's planning stage, I was convinced that this trust deficit must be addressed appropriately.

The presence of women SHGs in water service delivery and the improved quantity and quality of water supply have significantly contributed to reducing the trust deficit. Over 900 women from women SHGs have been inducted across 115 ULBs for doorstep services and manage an annual revenue collection of about ₹300 crore. This move has helped improve gender parity in urban service delivery and improved women's livelihood.

The community partnership model of engaging women for water service delivery is vital in sustaining water sector transformation and has received several national accolades, encouraging other states to adopt similar community-led models.

We come to the findings of the last category of the evaluation framework: sustainability and replicability. Usually, when we talk of sustainability, the focus is always on financial viability. This is undoubtedly important, but sustainability also depends upon several other aspects such as employee productivity, robust institutional mechanisms and sound policies and procedures.

One aspect that emerged strongly from the impact study was that the staff at different levels—managers, engineers, operators and Jal Sathis—took pride in their work. They felt that they were championing a cause, which I believe is fundamental to sustainability. Such motivated staff members are a vital cog in the wheel of sustainability. It is now up to the senior management to ensure that they remain motivated and can take on new emerging challenges.

Revenue collections have improved significantly with 100 per cent metering and the Jal Sathi-driven billing and collection process. For example, in Puri, revenue collection efficiency has improved from below 50 per cent to over 97 per cent, which is above the desired national benchmark of 90 per cent.

The behavioural changes observed in consumers have also been encouraging. Before DFT, citizens preferred their own groundwater supply over the public water supply system due to a lack of trust in the government-run water supply system.

The decision to leverage community partnership in the DFT Mission and engage women in service delivery has been a game changer that turned the table in favour of the public water supply system. If not for this partnership, regularizing illegal HCs and ensuring 100 per cent HCs would not have been possible.

The Jal Sathis went door to door and demonstrated water quality through on-site tests to gain the confidence of the public. Extensive IEC campaigns, especially the digital screens set up at strategic public places showing real-time water quality data, also helped build people's trust.

The shift to directly drinking from the tap is a significant change in behaviour. Additionally, compulsory metering of HCs and regular billing through Jal Sathis instilled discipline in water consumption, thereby contributing to system sustainability.

The need for continuous upgradation of skills led to the establishment of the OWA to train key functionaries, from operators to policymakers, through periodic capacity building.

The increased water supply interventions also enhanced the employability of the local population and migrant labourers. Local plumbers were trained and empanelled for HCs under the DFT Mission to prevent faulty connections and leakages. Many migrant workers who had returned to Odisha during the COVID-19 pandemic were also trained and provided with jobs under the mission.

Customer satisfaction through timely grievance redressal has been prioritized. The Centralized Customer Care Centre established under the DFT Mission with IVRS facility for automatic complaint lodging and transfer of complaints to the concerned staff for action enables online real-time tracking of grievance redressal.

Complaints are received through the 24×7 customer care centres and satisfactorily redressed within the prescribed time limit. Periodic meetings conducted by the field engineers with Jal Sathis are an opportunity for getting feedback on the prompt redressal of grievances and customers' satisfaction level.

The DFT in Puri has a notable influence across the country. Its successful implementation and branding have led to a demand for such water supply systems across cities in India. Some leading national-level political parties added DFT water in their election manifesto, hinting at the popularity of the concept.

Encouraged by the success of the DFT Mission in Puri, the Government of Odisha decided to upscale it to all 115 cities across the state in a phased manner. All major cities are covered in this phase, including Bhubaneswar with a population of 1.1 million, Cuttack with a population of 0.9 million and Berhampur with a population of 0.45 million. The completion of Phase 1 will see about 4 million people (60 per cent of the state's urban population), including 1 million slum dwellers, getting 24×7 DFT water in their homes.

Odisha's DFT Mission has become a lighthouse for the entire nation. Taking note of the success of Puri's DFT mission, the Government of India has mandated cities to provide drinking-quality water from taps in its AMRUT 2.0 guidelines which were launched on 1 October 2021. Several components of the DFT Mission find a place in the AMRUT 2.0 guidelines, including community partnership, metering and reduction of NRW as essential features.

The CPHEEO of the MoHUA has drafted and published

a case study of the transformational journey of Puri's DFT Mission, inspiring water utilities across India to take up such projects. So far, water utilities in over 20 states have sent their teams to Puri to learn from its DFT Mission, evincing keen interest in replicating the project in their states. The DFT Mission has set a new benchmark in drinking water service delivery in the country.

Before the mission, providing quality drinking water from taps to people in India seemed to be a distant dream. Such is no longer the case as the national and state task forces on 24×7 water supply created under AMRUT 2.0 met in Puri in October 2022 and committed to replicating the DFT model in more than 100 cities in various states. Presently, more than 600 DFT projects are at various stages of implementation in cities in several states under AMRUT 2.0. What can be more fulfilling for our team than seeing our DFT Mission carried to every nook and corner of the country where people deserve such services?

Odisha's DFT model has emerged as a beacon, inspiring other cities and states. They reached out to WATCO for providing end to end project management consulting services for the implementation of 24×7 DFT in their respective cities.

In a pioneering government-to-government collaboration, the Chennai Metropolitan Water Supply and Sewerage Board (CMWSSB), in the state of Tamil Nadu with a legacy of more than half a century, has sought consulting services of WATCO on a nomination basis (without tendering). The contract agreement was signed between CMWSSB and WATCO in July 2023 to provide project management consultancy (PMC) services for the implementation of the DFT project in Chennai. The ₹1,500 crore World Bank-supported project is estimated to benefit over a million people in Chennai's heartland, in the next four years.

Similarly, the Delhi Jal Board (DJB), a titan among water

utilities, not just in Asia but potentially one of the largest water utilities across the globe, servicing about 25 Million people, has sought the PMC services of WATCO to replicate Odisha's model of 24×7 DFT across Delhi, the capital city of India. With an agreement on the horizon, WATCO is poised to begin its transformative odyssey in January 2024, bringing 24×7 DFT to the vast expanse of Delhi. This ripple of change, emanating from Odisha, is reaching north and south, far and wide, as more states line up, eager to replicate this model in their states.

With such mega end-to-end PMC assignments in hand, the necessary institutional improvement including the PMC division of WATCO is poised to be strengthened.

It brings immense pleasure and a feeling of profound pride to see utilities as grand as CMWSSB and DJB seek inspiration and guidance from WATCO, a mere 5-year-old organization. It is a testament to Odisha's visionary leadership and the magic of putting people first in your thoughts and deeds in public service delivery.

17

24×7 WATER SUPPLY: MYTHS BROKEN

I can recollect my meeting in 2017 with one of my close friends who was then the urban secretary of a state that has done significant work in the urban water sector. By that time, we were seriously contemplating 24×7 DFT in Odisha, which was in the ideation stage.

When I sought his views on 24×7 water supply, he said, 'Mathi, never make the mistake of going for 24×7. It is neither feasible in our cities nor sustainable.' I was taken aback by his warning and decided not to share our idea of moving towards 24×7 DFT with him as I was scared of further discouragement.

While returning I made up my mind that we must initiate 24×7 DFT. It was a strong gut feeling and call from within but without rationale at that time. The mission was strongly driven by my passion backed by the strength of my team and the success tasted thus far. Above all, 24×7 DFT would take our journey to the highest level to realize the dream of CM Naveen Patnaik.

However, I was clear that we must dive deep into prior experiences within and outside the country to understand all the possible pitfalls before venturing into the implementation of 24×7 DFT. We started interacting with a few experts closely associated with 24×7 projects in India and abroad. During those interactions, we got to understand the instances and causes of failures. In fact, we gathered more knowledge on how to fail in the journey of 24×7 than how to succeed. In a way, they were very useful for us as we knew what not to do.

This understanding helped us draw up our strategies on how to discard some of the negative opinions associated with 24×7. I've always believed that if I want to achieve something, I have to strongly embrace a genuine cause.

Therefore, we started focusing on why we needed a 24×7 water supply and how we were in a good position to achieve it. We started counting our strengths and the successes achieved so far, which gave us the traction to look ahead and continue the journey with abundant caution.

In the process, we needed to let go many of our conventional understandings. Today, I believe that the team's process of unlearning and relearning was critical. Given all the discouraging stories about 24×7 water supply, we were cautious of failure from the beginning and carefully measured each of our steps.

In the successful implementation of the DFT Mission in Odisha, we have been able to dispel several myths surrounding the 24×7 water supply system. I believe it is important for me to elaborate on those myths for the benefit and understanding of other utilities and cities so that they will gain confidence in implementing a 24×7 DFT water supply.

Myth 1: Water demand significantly rises if you supply water 24×7, leading to water shortages.

This belief is based on the following apprehensions:

a) The demand will increase sharply as people will start using more water with 24×7 water availability at their premises.
b) Leakage points will release water 24×7 as compared with intermittent water supply where the leakage is limited to supply hours.
c) Due to an increase in demand, source sustainability

becomes difficult and the system will collapse and return to intermittent supply.

When we first completed and commissioned 24×7 DFT in Ishaneswar Slum, we indeed observed a steep increase in demand from the very first day. To tackle the sudden rise in demand, door-to-door communication was undertaken by Jal Sathis.

People were made aware that they need not store water as they would now get round-the-clock water whenever they opened their taps. Often, storage leads to wastage as people tend to store and then throw away the unused water the next day to store fresh water. The elimination of household storage helped reduce consumption and demand.

Further, individual household water meters and bulk flow meters installed in lanes were diligently observed, and leakage points were identified and repaired by deploying teams day and night. CCTV inspections inside pipes were carried out with the help of a start-up team from IIT Madras to help identify leakages Extensive communication regarding responsible water usage, metering and intensive identification of leakages and repair work helped stabilize the system, and the demand came down to normal levels within a month.

The lessons learnt from Ishaneswar Slum were implemented while upscaling, and we succeeded in stabilizing demand within one to two weeks after commissioning a zone.

Our experiences in 24 cities covering 2.5 million people revealed that the demand initially increased for a few months. Once metering was done and regular billing started, the system stabilized with the plugging of all leakage points.

The demand subsequently decreased to levels that existed before 24×7 and even lesser in some places. Effective leakage control, behavioural change due to effective communication and disciplined consumption due to metering were the key elements in busting the myth.

Myth 2: 24×7 water supply will lead to increased water wastage.

The belief is that people will resort to wasting water by consuming irresponsibly due to round-the-clock availability at their premises and low tariffs.

Our experiences revealed that with effective door-to-door communication by Jal Sathis, metering consumption and regular billing, even at the existing tariff rates, people started consuming responsibly.

The reliability of supply and consistency of pressure and quality raised public trust and confidence, resulting in changes in consumers' behavioural patterns.

Myth 3: Need for high capital cost.

It is believed that implementing a 24×7 water supply project involves a high capital cost due to the requirement of completely replacing existing water supply infrastructures. This is because in the case of a PPP or EPC, the concessionaire or contractor resorts to full-scale replacement of all existing infrastructure to eliminate the risk of performance failures or avoid frequent replacements during their O&M period.

Generally, bidders do not take up detailed surveys and assessments of the existing infrastructure while bidding, even though usually three to four months are given for a thorough assessment of the existing infrastructure.

In our case, we did a thorough assessment of the existing infrastructure and identified the gaps to be filled for 24×7 water supply without depending on the contractors. Hence, the scope of work was clearly outlined without any ambiguity or the possibility of bidders adding to the cost to meet unforeseen risks.

When we reviewed the capital cost of previously implemented and the ongoing 24×7 water supply projects under construction across the country, the cost seemed to be exorbitant. It was also noted that most of the projects were tied up with foreign loan funding. The high capital cost of all these projects gives the impression that the capital cost for the implementation of 24×7 water supply projects is prohibitively high.

While surveying and assessing the condition of existing infrastructure, we found that most of our infrastructure, such as pipelines and WTPs, was not very old as they were constructed during the JNNURM and AMRUT 1.0 between 2012 and 2017.

Therefore, we upgraded some of the old infrastructure and filled the gaps to convert the existing intermittent water supply system to a 24×7 system without resorting to wholesale replacement. Thus, our capital cost worked out to be reasonable, not exorbitant.

The secret to implementing a 24×7 system at a reasonable cost is assuming responsibility for the survey, condition assessment and infrastructure gap finding; in-house planning by the water supply authorities and entrusting only the construction responsibilities rather than the survey, assessment and scoping responsibilities to the contractor who is an outsider with no knowledge of the existing infrastructure. This also ensures that 24×7 water supply projects are executed without depending on international funding agencies.

Myth 4: High operation and maintenance costs make 24×7 water supply unaffordable and unsustainable.

This is based on the following assumptions:

a) 24×7 water supply will cause 24×7 leakages and water wastage and lead to a disproportionately high demand for water. This would lead to increased pumping requirements and a high energy cost.

b) Operating a 24×7 water supply will require a huge workforce and result in high human resources (HR)

> costs. Due to high operational costs, cities may be unable to financially sustain the system, thereby leading to the collapse of the 24×7 water supply.

The myth related to an increase in water demand and increased water losses has already been explained *(refer to Myth 1)*.

Our experience revealed that though energy consumption and cost increased initially, after the reduction of water losses from around 50 per cent to nearly 15 per cent, the water demand and energy consumption reduced significantly.

The deployment of an industrial IoT-based smart water management system for O&M reduced the requirement of technical manpower. For the increased activities related to services at consumers' doorstep, such as meter reading, bill generation, water tax collection and grievance redressal, partnering with Jal Sathis eliminated the need for an additional workforce.

As Jal Sathis work on a performance-linked incentive basis, we were able to optimize the outputs and costs. Hence, our experience revealed that there has not been an unreasonable increase in operational costs.

Myth 5: Huge requirement for manpower.

It is generally believed that 24×7 water supply will require more manpower than conventional water supply systems. It will require the recruitment of many engineers and support staff for meter reading, billing, collection, etc., which is generally not easy in government systems. Furthermore, approval by the finance department, the limitation of cadre restructuring and a long-drawn recruitment process remain challenging aspects.

With community partnership accompanied by technological interventions, we created a sizeable grassroots-level workforce fairly quickly without having to go for large-scale government staff augmentation.

Myth 6: High-end consultants are a prerequisite.

A common assumption is that high-end consultants are essential to design and execute 24×7 water supply projects. In most of these projects, including ongoing ones, designing and engineering have been entrusted to high-end experts or consultants, many of whom happen to be foreigners.

In contrast, our projects are designed and executed by our own engineers by engaging local contractors without onboarding high-end experts and consultants.

During the pilot phase, we consulted a few experts to understand the technical aspects; thereafter, we took up extensive capacity-building of in-house engineers, and the upscaling process in the cities is now being done by our own engineers.

I feel that with the experience gained during the implementation of 24×7 DFT projects, each one of my team members has now become an expert in planning, designing, executing and operating 24×7 water supply systems.

Myth 7: Revenue recovery is impossible.

Many cautioned us that after such huge investments, revenue recovery would remain a huge challenge in 24×7 water supply. Revenue recovery depends upon customer satisfaction and willingness to pay, which are closely linked to the quality of services provided.

I believe that if service quality is good, customers will be willing to pay. That's why we first focused on improving the services in terms of quantity and quality, followed by metering the connections.

The return on investments for drinking water supply cannot and should not be measured only in terms of financial returns.

The economic returns of such investments in terms of improved public health indicators and enhanced economic productivity of citizens are difficult to quantify but are equally important. Therefore, we looked at return on investment in financial and economic terms.

With the provision of round-the-clock DFT-quality water, consumer doorstep service delivery by Jal Sathis and improved public trust and confidence through measures such as a digital display of real-time water quality data in public places, deployment of labs on wheels and quick response team, our revenue collection efficiency in Puri increased to 97 per cent with 100 per cent metering of HCs.

Myth 8: Community partnership in such a technically complex sector is likely to backfire.

We decided to go for community partnership at the grassroots level by engaging local women from marginalized sections to manage public water supply distribution. However, many feared that involving less-educated women in consumer relationship management for such a technical subject as water supply was a huge risk and would fall apart.

We experienced that challenges like illegal connections, tampering with water meters, non-payment of water charges, etc., were not easy to tackle for government officials. I strongly believe that if adequately incentivized, women SHG members who hailed from the same locality could tackle these issues effectively with the support of their local community.

With intensive training, well-defined standard operating procedures and a performance-linked incentive system, our Jal Sathi model is now recognized as a game changer and an inspiration for other states.

TO BELIEVE IS THE FIRST STEP TO SUCCESS

If you attempt to do any good work, people will often say with cynicism, 'No. It is not possible in India. There is large-scale corruption. There is a contractor–politician–bureaucrat nexus. People are bad, and they won't pay for any service. They want everything for free. The poor will not pay. The private water business lobby won't allow such a project to succeed. The water tanker business lobby will sabotage such a venture.'

Our answer to all this is that Odisha has achieved this in 24 cities and that too in three years. If this can be achieved in Odisha, why not in other places?

THE ESSENCE

The greatest enemy of knowledge is not ignorance; it is the illusion of knowledge. Breaking myths is the key to unlocking true knowledge.

—DANIEL BOORSTIN

Many times, myths and misconceptions hinder our ability to pursue innovative solutions and achieve great results. In the face of cynicism and doubt, we refused to back down from our vision of implementing 24×7 DFT in Odisha. In the process, we shattered many myths surrounding this idea. It proved that an increase in demand could be tackled through effective communication to encourage responsible consumption and that water losses could be reduced through the need-based upgradation of existing infrastructure and leakage control.

Most importantly, we proved that an investment in water is investment in preventive public health, which is essential to achieving long-term economic returns. Our journey is a shining example of how breaking myths and pursuing innovative ideas can lead to real progress and positive change.

The successful implementation of impactful programmes can create public goodwill, which is crucial in neutralizing negativity and cynicism surrounding public service delivery. Today, Odisha's DFT is not a dream but a reality. It's a game changer that's bringing safe, clean water to millions of people.

Breaking myths: the key to progress and positive change.

ACKNOWLEDGEMENTS

I extend my profound and foremost gratitude to Shri Naveen Patnaik, CM of Odisha, and Shri V.K. Pandian, 5T Chairman, for entrusting me with the opportunity to translate their vision into reality. Without their unwavering support, I would not have had the privilege of authoring a book like this. By providing a strong foundation and acting as the driving force behind the Odisha model of 'Drink from Tap' (DFT), they have gifted the people of Odisha one of the greatest and most innovative projects in the country's water sector, the benefits of which are now transcending state boundaries.

My sincere thanks to Mr P.K. Swain without whom I would not have dared to embark on the daunting task of translating the DFT dream into reality. Though I always take pride in proclaiming that it was me who discovered a great technocrat in P.K. Swain, the fact remains that it is my blessing that he is with me, playing a pivotal role in establishing the DFT model.

Next on my list is Chinmay Tripathy who has been travelling with me throughout the DFT journey, right from its conceptualization. His contributions in scripting the DFT model, as well as this book, have been absolutely unmatchable. In my experience of working with several teams in the implementation of various transformative initiatives like the Jaga Mission, Faecal Sludge and Septage Management Model, etc., I've not met anyone coming close to Chinmay's contributions and sincerity.

Writing a book is not my cup of tea. It was at the

insistence of Bala Sir [Shri R. Bala Krishnan, IAS (R)], my senior colleague in service and cadre, and more of an elder brother, that I should record my experiences for the benefit of the water sector. This gave me the impetus to pen my experiences to bring out this book. My heartfelt thanks to him.

I am grateful to the Hon'ble Ministers, Chief Secretaries, my colleagues in PHEO and WATCO and the great Team Odisha who have reposed faith in me and supported me all along.

My sincere thanks to the ground brigade of engineers and officials of WATCO for their unwavering dedication in contributing to the success of DFT. Special thanks to our community partners (Jal Sathis), for playing a pivotal role in DFT's success. Heartfelt appreciation goes out to the millions of urban Odisha water supply consumers who have actively cooperated through the entire process. *People First* is a testament to the collaborative synergy of all these key players.

My earnest thanks to Ms Nalini Elumalai and Mr Eklavya for slogging for days and month, helping me fine-tune this book through repeated joint virtual sessions.

I am eternally grateful to many others whom I won't be able to list here due to space constraints.

Last but not least, I feel highly indebted to my parents for laying the foundation that shaped me and instilling the values that guide me. My family has been my pillar of support without whom I would not have reached where I am now.

LIST OF ABBREVIATIONS

ABD	Area-based Development
AMRUT	Atal Mission for Rejuvenation and Urban Transformation
ASCI	Administrative Staff College of India
ATM	Anytime Water Machine
AWSSAR	Association of Water Supply and Sanitation Organization
BASUDHA	Buxi Jagabandhu Assured Water Supply to Habitations
BBWSP	Bhubaneswar Bulk Water Supply Project
BMC	Bhubaneswar Municipal Corporation
CEO	Chief Executive Officer
CM	Chief Minister
COO	Chief Operating Officer
CPHEEO	Central Public Health and Environmental Engineering Organization
CSR	Corporate Social Responsibility
CWIS	Citywide Inclusive Sanitation

DFT	Drink from Tap
DMA	District Metering Area
DMF	District Mineral Fund
DPR	Detailed Project Report
EE	Executive Engineer
EIC	Engineer-in-Chief
EO	Executive Officer
EPC	Engineering, Procurement, Construction
ESR	Elevated Storage Reservoir
FC	Finance Commission
FIR	First Information Report
GBWSP	Greater Berhampur Water Supply Project
GIS	Geographic Information System
HC	House Connection
H&FW	Health and Family Welfare
HUDCO	Housing and Urban Development Corporation Limited
H&UDD	Housing and Urban Development Department
IAS	Indian Administrative Service
IDCO	Odisha Industrial Infrastructure Development Corporation
IDSP	Integrated Disease Surveillance Programme

IEC	Information, Education and Communication
IIT	Indian Institute of Technology
IoT	Internet of Things
IRC	International Water and Sanitation Centre
IVRS	Interactive Voice Response System
JICA	Japan International Cooperation Agency
JNNURM	Jawaharlal Nehru National Urban Renewal Mission
KYC	Know Your Customer
LIG	Low-income Group
LPCD	Litres per Capita per Day
MD	Managing Director
MIG	Middle-income Group
MLA	Member of Legislative Assembly
MLD	Million Litres per Day
MLALAD	Member of Legislative Assembly Local Area Development Scheme
MoHUA	Ministry of Housing and Urban Affairs
MoU	Memorandum of Understanding
MP	Member of Parliament
MPLAD	Members of Parliament Local Area Development Division
MPoS	Mobile Point of Sale

MSG	Mission Shakti Group
NDMC	New Delhi Municipal Corporation
NISER	National Institute of Science Education and Research
NIUA	National Institute of Urban Affairs
NRW	Non-revenue Water
NULM	National Urban Livelihood Mission
ODF	Open Defecation Free
OMBADC	Odisha Mineral Bearing Areas Development Corporation
O&M	Operation and Maintenance
OSD	Officer on Special Duty
OWA	Odisha Water Academy
PHEO	Public Health Engineering Organization
PIL	Public Interest Litigation
PLC	Programmable Logic Controllers
PMAY-U	Pradhan Mantri Awas Yojana-Urban
PPMS	Project Performance Monitoring System
PPP	Public–Private Partnership
PR&DW	Panchayati Raj and Drinking Water
PSP	Public Stand Post
PUB	Public Utilities Board

PVC	Polyvinyl Chloride
PWD	Public Works Department
RDC	Revenue Divisional Commissioner
RFID	Radio Frequency Identification
SAAP	State Annual Action Plan
SCM	Smart Cities Mission
SDG	Sustainable Development Goal
SE	Superintending Engineer
SHG	Self-help Group
SOP	Standard Operating Procedure
SUDA	State Urban Development Agency
SWA	Singapore Water Academy
SWM	Solid Waste Management
TPMA	Third-party Monitoring Agency
TWAD	Tamil Nadu Water Supply and Drainage Board
UFW	Unaccounted for Water
UIDSSMT	Urban Infrastructure Development Scheme for Small and Medium Towns
ULB UN	Urban Local Body United Nations
UNICEF	United Nations Children's Fund
USP	Unique Selling Point

VGF	Viability Gap Funding
WATCO	Water Corporation of Odisha
WATSAN	Water and Sanitation

16

HARVESTING DREAMS

Nothing is more satisfying than seeing disadvantaged communities benefiting from government schemes. Unlike other sections of society, these communities typically rely solely on government interventions to improve their quality of life.

A key challenge in developing countries like India is translating political vision into ground reality. However, today, 97 per cent of urban households in Odisha have piped water connections, with 105 out of 115 cities achieving 100 per cent HCs. Before June 20[illegible], all households in urban Odisha will have piped water connections.

Moreover, out of around one million water supply connections across the state, of these about 4,00,000 connections are in slums, benefiting the urban poor. Water connections in slum households were earlier considered not permissible due to land ownership issues until we announced a 'leaving no one behind [illegible]' policy.

Before the [illegible] water transformation in Odisha, the poor [illegible] had intermittent or no access to piped water and were exposed to consuming water from other potentially contaminated sources. This resulted in many cases of waterborne diseases every year.

The lack of access to piped water supply at home disproportionately affected the women and children. They were compelled to spend hours fetching water when they